DUTY
HONOR
PURPOSE

THE HISTORY, THE HEROES, & THE HUMANITY OF MILITARY SERVICE

★★★★

CHARLES ODONNELL

This book is dedicated to our nation's defenders; past, present, and future.

If you are a veteran, this book is a Tribute to your past service.

Charles F. Tiessen - U.S.A.F.	[R.I.P.]
James F. Hackett - U.S.M.C.	[R.I.P.]
John C. Stewart - U.S.M.C.	[R.I.P.]
Martin D. Wright - U.S. Army	[R.I.P.]
William H. Foster - U.S. Army	[R.I.P.]
Joseph A. O'Donnell - U.S. Army	[R.I.P.]
James B. O'Donnell - U.S. Navy	[R.I.P.]
Charles P. O'Donnell - U.S. Navy	[R.I.P.]
Gerald J. Marklin - U.S. Navy	[R.I.P.]
Richard E. Foster - U.S. Navy	[R.I.P.]

Special Thanks to the following:

Thomas Phiel / Michie Saint USN
Edward Beach U.S. Navy Captain (retired)
Cecil Stellyes U.S. Navy Commander (retired)

WARNING!

This book could change your life!

Preface

This is a book about military service and the unique aspects of military service that turn teenage boys and girls into responsible, courageous, and extraordinary men and women.

While the book touches on the unique nature of each service and how they are organized, the more interesting elements are the extraordinary skill and courage demonstrated by kids, who just a year ago were struggling with their high school homework

Military service changes people. It makes them responsible. It makes them respectful. It reinforces their faith, their values, and their humanity. I should know, I was one of those kids.

In 1965, the Viet Nam war was moving into high gear. My high school friends were either enlisting in the military or getting drafted. I soon realized that If I did nothing, my fate and my future life was going to be decided by the government instead of me. This pushed me into really thinking about what I wanted to do with my life, and how could I take control of what seemed to be uncontrollable. My friends and I anguished over our futures.

Each of us chose a different path. One friend chose the Army. One friend chose the Navy, two of my friends chose the Marines, and I chose the Air Force. ……….. We all survived.

This book is for every young person who is undecided about what they want to do in life, whether that is to build a solid career, raise a family, and lead a good, safe, and productive life, or simply drift from one momentary opportunity, to the next.

You are not alone. Some of the entertainment industry's biggest names were once teenagers contemplating some of the same apprehensions. In this book, you will read about James Stewart, Charles Bronson, Lee Marvin, Julia Childs, Marlene Deitrich, and Bea Arthur, all of whom played key military roles during WW II.

You will also read about some young people whose names are not well known; like Guy Gabaldson, who convinced a division of Japanese soldiers to surrender to him, and then single-handedly marched 1300 enemy soldiers into his unit's camp to surrender.

I wrote this book for the young men and women who may be wondering. if this is something you really *want* to do, or is it something you really *need* to do? When all is said and done, it is only a couple of years out of an entire lifetime, and I assure you, it will change your life. Who knows, you may even like it!

If you are a young person and did not finish your education, are unemployed, and/or have no skills, your efforts to become a productive citizen, earning a good living, is going to be a continual challenge. There is an alternative. If you want a preview of what the military is all about, read on.

* * * * *

Table of Contents

Introduction

What is it about our military that is so frightening? Is it the guns? Is it the thought of being killed? If you are between 18 and 28 and that is your perception of the military, then you are probably missing out on one of the great opportunities of your life-time. This is not a recruiting pitch; it is merely a suggestion that you carefully think about and consider what you may be dismissing.

If you are currently unemployed, and have been for some time, you probably don't realize that you could be working at a job that will teach you a trade, provide you food and lodging, and pay you $25,000 a year to start. If you make a career out of it, a six-figure income is possible, if you are willing to work for it.

If you are a high school student and would like to go to College, but can't afford it, there is an alternative to giving up on your dreams. If you join the military, you can finish your education in the service, and they will pay YOU.

If you are someone who has few friends, has few hobbies, or is just plain bored, and you looking for direction and purpose, the military can provide both, or all of the above, if that is what you are looking for. You can become part of an elite group that is respected and admired. Some of the services take people up to age 42.

If any of this, piques your interest, go talk to a recruiter. It costs you nothing but your time. If that is intimidating, go talk to a veteran. Almost every veteran will tell you that the military changed their life, and if asked.... they would do it again.

This book is about the military experience. It tells you how the military is organized what each service does, and a glimpse of what our Hollywood celebrities did when they proudly served our country. It touches on the attributes of military service, and the unsung heroes whose names have been lost to history.

Our military promotes family values in a big way. Yet this side of the military is largely unknown. The friends you make in the military may even become part of your extended family. You might even meet your future spouse while serving in the military. Our military is not something to be feared, but instead something to be cheered. The military protects our families *and*. our freedoms.

Freedom is not free, and as a result, we have found it a necessity to maintain a large, capable military, to defend our borders, as well as defend our values.

Unlike many countries, where military service is mandatory, our military is made up of volunteers. We rely on citizens who are willing to step up, and commit some portion of their working life to maintaining the freedom that we all enjoy and benefit from. To those that choose to do so, we thank you, and we honor your commitment to preserving our way of life.

The decision to serve one's country is indeed a personal demonstration of duty, honor, and purpose. It is our obligation, to support the society that provides us religious freedom, economic opportunity, and the life principles of doing the right thing, at the right time, for the right reason.

* * * * *

Chapter One

Living life to the fullest

They say that the most important decision of your life, is when you decide to get married, which will control and influence virtually all of your life's decisions for as long as you live.

The second most important decision that you will make is whether to serve your country by enlisting in the military. This is also a decision that can impact you for as long as you live.

One thing you will learn while in the military, is that your perspective on life, and the friends that you decide to associate with, will change you dramatically, as you quickly learn who you can rely on, who you can trust, and why that is important. Picking your friends in the military can literally save your life.

Experiencing life, and living it to the fullest, is a common theme of young adults. But what does that really mean?

If it means experiencing different emotions, visiting other countries, or living amongst other cultures, then maybe the military really is the one element you are searching for.

Bungee jumping, while tied to an 80-foot elastic cord may be momentarily thrilling, but so is jumping out of a plane at 10,000 feet.

Being a Lifeguard at the beach can be fun one moment, and harrowing the next. You want thrilling? Try dangling from a Coast Guard Rescue helicopter above 20-foot waves in a 45 MPH wind while 200 miles out at sea.

Think you are brave? Try investigating an enemies' dark underground tunnel in a far-away place with a hand grenade in one hand and only a pistol in the other.

Do you want to be more focused? How about cruising in the Red Sea on a nice sunny day in 100-degree heat, worried about enemy drones, missiles, submarines, or suicide pilots? That should help you sleep better, knowing that you are the prime target of every adversarial weapon available within 500 miles, trying to catch you at a lax moment, like when you are sleeping.

If you actually want to live life to the fullest, the Armed Forces can accommodate you at any level of risk you desire.

"Being all you can be" may be a recruiting slogan, but it is a subtle reminder that we are not invincible, and if you truly want to experience everything life has to offer, military service will certainly set you off on the right foot and expose you to the traditions and cultures of other countries, as well as meet some of the most interesting people of your life.

On the other hand, if you don't see yourself as wanting to push the life envelope, how about taxing your brain power to help decipher spy photographs, becoming an expert sonar interpreter, learning to operate sophisticated cipher technology, launch a moon rocket, or simply becoming a jet mechanic or an electronics expert. Learning a skill that you can use later in civilian life, is invaluable.

Deciding whether or not to join the military is the second most important decision you will likely make in your life

Your Career

One career option is to join the military, as repulsive as that idea may seem to some. The fact is, that if you are one of those turned off by the thought of military service, it is probably because you know nothing about it. Your fears are unfounded.

It always amazes me how quickly young men and women dismiss the thought of military service. You may think the military is what you see in the movies, lots of death and destruction coupled with needless loss of your liberty, your limb(s), or your life. If that is your mindset, perhaps you better read this book in its entirety, and think about what is important to you. Not because joining the military is dangerous, or even risky, but because the experience will change who you are.

Some join the military for the adventure, some to escape their surroundings, and some because instinctively, they just know it is the right thing to do, for themselves, as well as their country.

Almost no one talks about the vocational training opportunities that are available to service members. It does not matter if you have a specific career in mind or not. It does not matter if you are a college graduate (although that is a plus). What does matter, is your future.

If you do not know what you want to do, the military has one of the largest selections of career options anywhere. Ninety percent of the job classifications in the military are non-combat positions.

That does not mean that you won't be taught how to defend yourself, or operate a weapons system in an emergency. But on a daily basis, you will be learning a skill., not shooting at anyone.

Most importantly, if you can pass the Armed Service Vocational Aptitude Battery (ASVAB), you can choose the career of your choice. A college degree is not a pre-requisite for many positions. The test will indicate what choices you will have the best chance of success at.

After your vocational training has been completed, you may find that you are better suited for another type of work. The military even has a process by which they can allow you to do that, provided you can meet the requirements, and have the recommendation of your supervisor. This process is much easier than trying to change careers in civilian life, and you experience no period of unemployment. You can even complete your college degree while in the service.

There are very few civilian jobs that do not have an equivalent vocation in the military. These jobs are needed to keep our military functional, dependable, & proficient, and are the backbone of our military services. Choosing to get your training courtesy of the military is free, but it does require a service obligation on your part.

Joining the military is not an easy decision to make, but it is one that you will likely never regret. Military service for all its detractors, is still one of the most positive life-changing events you will ever experience. Ask any veteran. Virtually, all would do it again.

As it turned out, when I originally talked to a recruiter, I found out I had more choices than I ever imagined. I made up my mind that I would gather as much information as I could, and only then, after giving myself time to think about my options, would I choose an option that best suited my interests. The best part was that this was my decision.

My father served in the Army in WWII, his two brothers served in the Navy, and my older cousins also served in the

Navy during the Korean war, as did both my future brothers-in-law, so military service in and of itself, was not the scary part. It was not knowing where I would be sent, what skills I would be taught, and putting my own career on hold. Could I survive the ordeal? I would also be living with total strangers, and who knows what they might be like. As it happened, we all shared the same anxiety

At that point in my life, as a single guy, this was the most monumental decision I would make about my life. No one else would be making this decision for me, this was my decision, and mine alone. Now I had to find out what I was made of.

I had tried to join the regular Air Force, but with Viet Nam underway, there were no openings in the career fields I wanted, so I joined the Air Force Reserves. This turned out to be a great option, & what I learned, made me feel even more at ease with my decision.

I preferred to be in control of my own life, and the Reserves gave me a taste of military service while providing me a potential future career direction & a military service perspective that I never thought about. It also provided extra income that came in handy later on, and it led to a job with a major airline when I returned home.

Teamwork

There is one more element of military service that should be mentioned here; Teamwork. The ability to accomplish things through teamwork cannot be understated. Learning to work with others, to depend on others, and knowing that they depend on you as well, is a great confidence builder, and a nice addition to anyone's personality. [As a person who worked in

Supervisory and Management roles during my entire career, I can tell you that ordinary people, that are properly directed and confidently led, can and do, accomplish extraordinary things].

As a Store Manager, an Operations Manager, a Logistics Director, and ultimately as a Chief Executive, I can tell you that your employees can make the best company fail, or the worst company succeed. It is that persistence, that ingenuity, and the teamwork of everyone working together towards a common goal, that often determines the difference.

Teamwork is more than just camaraderie. It is more than just being buddies. It is the mutual respect and loyalty and trust you place in others to help you accomplish difficult tasks. Ask a Navy Seal.

Discipline

You may not think much about discipline, but after your military career is over, you will likely seek civilian employment. If you are a former military man or woman, you will find that the discipline and the organizational skills you learned in the military, now provide you job preference for supervisory and management positions. Both of which, require leadership, teamwork, and organizational skills, just like you were taught in the military.

That screaming in your face during boot camp taught you discipline and consequences. Those physical training exercises taught you team work, and all those paperwork processes you tolerated in the military make you now realize that when you document things in an organized manner, it makes the follow-up easier, and provides accountability. When you start out in the military, it all seems like unnecessary bureaucracy, but in civilian life, you soon realize that the military methods were concise, accurate, and organized. You can go to any other military installation, and every process operates within the same parameters.

Empowerment

Once you are trained in a skill in the military, you will find that you are empowered to carry out your part of the military mission using your own initiative and your own judgement. There is not time for micro-management. You are trusted to take the initiative, and use whatever resources you have available, to accomplish your assigned task. No hand-holding. Your unit is depending on you, just as you are depending on them. Together, it works!

Your Retirement

Regardless of whether you enlist for a single term, or whether you enlist for twenty years or more, someday, you are going to retire. Your life will gradually change; physically, mentally, & economically. Those that chose to serve in the military as a young person may find that fifty years later, they can still reap the benefits of military service from some of the most unlikely places. You should plan to live a fulfilling retirement just as you plan to live a fulfilling working life.

Former members of the armed forces find that in retirement, they are afforded veteran discounts on travel, food, lodging, insurance and memberships. You may scoff at a 10% discount today, but when you are suddenly living on a fixed income after retirement, every discount leaves you more spendable income. Simply saving small amounts such as a discount on a dinner, an admission, or a purchase at Lowes, Target, or AAA travel insurance can add up to $50.00 a week or more than $2500.00 in a year, and that is a very conservative amount.

Those enlisting after 2018 can participate in the Thrift Savings Plan. TSP is the military version of a 401 Plan. After serving two years, you may qualify for government matching funds up to 4%.

After 20 years, you qualify for a full retirement, which is 40% of your highest three years. Each year served after 20 adds 2% more per year.

If you served twenty years or more, there are a lot more benefits that you now qualify for. Items such as access to free veteran medical services, including vision and dental. Access to the Base Exchange network for sizeable retail discounts. Add to that, low interest veteran mortgage loans, and auto insurance. Plus, you saved on all those free and/or discounted services. Want to go to college after retirement? The government may pay for that as well.

But wait, there is one more thing you may not have even considered; If you enlisted when you were 18, and you serve for twenty years, you will only be 38 or 39 when you retire from the military. If you work for a private employer for your remaining working life, you will now qualify for a military pension, a private employer retirement, plus Social Security. That means 3 paychecks to help you live that fulfilling life you are working so hard to achieve.

Want even more? If by chance, you marry another member of the armed forces, who also served twenty years, & then they also worked for a private employer for the remainder of their working life, you could both be receiving multiple retirement checks.

Still not convinced? Try this on for size; In addition to the above scenario, you purchased a home with a VA loan. Upon retirement, you both decide to downsize. You sell the home, put $200,000 of your profit into a retirement annuity, and invest the rest. If your home was paid off, and you made $400,000.00 or more, after selling commissions, and you both invest the same amount with your profit, you could be receiving four retirement checks each, and still have some funds in a savings or investment account that pays dividends, while preserving your principal. That could net each of you extra money to leave your children. Not bad, eh?

* * * * *

Chapter Two

Military Life

As one who served six (6) years in the service of our country, I find that this notion, that we are assertively engaged in training kids to aggressively kill people is almost laughable. So, let me set you straight on a few key points:

1. In my six years of service, I was never issued a service weapon. I was taught to properly clean, maintain, aim, and fire a weapon, in order to defend myself, or others, but as for actually being issued a personal firearm, it did not happen. I was also never issued a combat helmet. That may seem surprising, but only about 10% of our armed forces are combat troops. Ninety percent (90%) of military personnel are in administrative or logistical support roles.

The majority of my time in the military was spent learning to become proficient in my assigned military vocation [albeit mechanic, parachute packing, fireman, or electronic technician, etc.] for the purpose of becoming a useful contributor to society, in or out, of the military. I learned to be organized and to follow directions. I learned the importance of making informed decisions. I learned to plan and anticipate, rather than becoming a victim of my own circumstances and carelessness. I also learned to be accountable for my actions. No excuses. Military training is focused on the mental aspects as well as the physical aspects of preparing for, supporting, and winning a conflict by everyone doing their job.

2. Women are making up more and more of our military. Currently, almost 18%, of our armed forces are women. Very few are in combat-related positions, but that is changing and it is only a matter of time before more women are permitted to be assigned to combat-related roles. We already have female pilots, female naval officers, and women army officers. They have proven themselves more than capable time and time again.

3. My basic training focused on learning how the military functions and how to take care of yourself, both physically and mentally. Becoming a responsible and disciplined team leader was the order of the day. In the case of the Army and Marines, they do get advanced training in combat strategy and techniques, but these represent a very small number of military troops. This is becoming increasingly so, as more and more automated and technological weapon systems, are developed. and introduced into the world's battlefields.

4. Responsibility, accountability, & attention to details were the primary life lessons that were focused on during my military training, and knowing the consequences when delinquency, dereliction of duty, and/or dishonesty prevailed. The military gives you purpose, direction, dignity, respect,

and recognition, something many youths strive for in civilian life.

5. Life on a military base is safer, and more secure than living off base, primarily due to the consequences that are spelled out in the Uniform Code of Military Justice (UCMJ). Loss of rank, Loss of pay, Forfeiture of Leave, and/or restriction to the base, are some of the more common penalties. Despite the highest firearms per capita of any city; military bases have almost no crime involving firearms, no violent crime, and one of the lowest crime rates of any population or organization. You are unlikely to find graffiti, bullying, or sex crimes and if they do occur, the penalties will land you in the brig for an extended period of time. No parole, no community service, and no slap on the wrist. Military discipline is unambiguous, it is swift, and it is harsh.

6. The Leadership and Teamwork attributes of military service will change your life. You will become more outgoing, more disciplined, and more patient and in control of yourself, than you ever imagined. You will quickly learn how to be more professional, more proficient, more organized, confident in your thinking and more controlled in your actions.

   When you leave the military, you will have the skills to compete for any job at any company with the confidence of knowing your military skills are valued.

7. The Military takes care of its own, from the time you enlist, to the time you retire. For individuals that remain in the military long enough to earn a retirement, the military continues to provide medical support, access to military

facilities, and even burial in a military cemetery upon your eventual demise.

For those that choose to leave the military, you leave knowing what sacrifice, commitment, and freedom are all about. For that, you receive a lifetime of respect for your service & a lifetime of experience that will shape your life.

In basic training, you are encouraged to write home regularly, so your parents would know you were all right, and to lessen their anxiety at home. We were told it was not un-manly to let your family or siblings know that you loved them. Family values are a part of military life.

We were encouraged to attend the weekly religious services of our choosing. We were reminded that there are no atheists in times of war. Having faith is often your strength. During the Viet Nam war, our American P.O.W.'s attested to this over and over. It was their faith that helped them endure.

You were taught how to do your own laundry, iron your own clothes, shine your shoes, brush your teeth, and shower regularly. You learned good personal hygiene, and proper etiquette. You were expected to maintain your personal belongings in a clean and orderly manner. In short, you were learning to become a responsible adult, not an aggressive "Rambo" killing machine like many people think.

You also learned that you were now under the legal umbrella of the U.C.M.J. The **U**niform **C**ode of **M**ilitary **J**ustice, which governs all military actions and establishes penalties for violating military rules, conduct, or orders.

Now lest you think all of this sounds too warm & fuzzy, rest assured, when you are in Basic Training, you will get your share of verbal abuse and criticism. This is done to get you to pay attention to the details, which can potentially save your life.

When basic training is over, you proceed to advanced technical training, where you are taught a skill and also the basics of defending yourself. How to properly clean and aim your weapon. Learning tactics that are currently being used by any opposing forces. But the non-technical focus is always on safety, and saving your own life and/or the life of your fellow servicemen, not on killing others*. As one Sergeant reminded us; *"Don't worry, when someone shoots at you, you will instinctively, know what to do, and why, so, pay attention".*

* With some exception for those small, elite, Special Operations teams that are focused on removing life threatening obstacles to survival, freedom, or victory,

---

In some corners of the world, there are large populations that resent the United States' way of life, and are determined to destroy it. Drug cartels, religious zealots, political extremists, foreign terrorists, and rogue regimes are all intent on weakening our resolve.

Some of this is based on economics and greed, but most of it is based on an extreme religious belief that portrays all God-fearing people as Infidels, who would strip *them*, of their power. They believe that God will reward them in heaven for dispensing with anyone like us, who believes differently. We are seen as the evil force in their religious universe. <u>Death to them, is not a sacrifice, it is the ultimate reward, so their tactics are neither logical, nor predictable</u>. <u>Their glory is in dying, not living</u>. That is why suicide bombers are martyred.

They are taught from birth, that they must initiate Armageddon to destroy a world controlled by Infidels. That is their purpose in life. Martyrdom, or God's glory, is achieved only after death. These extreme views are shared by millions of religious extremists, and are directed at all non-believers, particularly Jews and Christians. If you are an American or a European, you are considered fair game, regardless of your personal beliefs. That is the reality of our world.

Minorities and the Military

When you enter the military, you will learn that discrimination is not tolerated, and to reinforce that commitment, all recruits get the same haircut, the same uniform and the same pay. No one is assigned to any chore based on their race, color or creed. Jobs are assigned by Rank. If you are a recruit, you clean.

Every military person shares equally in temporary assignment to chores like K.P. duty, restroom sanitation, and litter patrol. You are assigned to a barracks based only on your rank or training status. Promotions are based upon test scores, demonstrated proficiency, and time-in-grade. Everyone who is the same rank, gets paid the same, eats the same food, gets the same housing, and are offered the same training options. Success in the military is the result of your personal effort, your achieved rank, plus your time in service.

These equalizing factors ensure that soldiers get to know and understand each other, trust one another, and work as a team. There is no time or tolerance, for jealousy or favoritism. This does not mean that people of the same background can't hang out together on their own time when they are off-duty, but active duty leaves little time for discriminating against a fellow soldier, male or female, black or white, Catholic or Jew.

For many young men, or women, this is the first time in their life that they had to trust someone of a different culture, or have someone of a different culture trust them. In the military, any serviceman can be counted on to help you, without hesitation. You are all one team focused on a single mission. Together everyone gets through this experience peacefully, proudly, and alive.

Women in the Military

In the past, the policy of the military services has been to avoid sending women into combat roles. However, that policy is gradually changing, Today, we do have women serving in the U.S. Air Force and Navy as Fighter Pilots, serving on ships as Captains, an in other vulnerable positions. The U.S. Space Force employs the most women. Gradually, some high-risk positions now include women..

The reason is simple; the military realizes that women in almost every situation are just as capable as men. The remaining obstacle is that our military has a high resistance to putting females in high-risk positions subject to capture, torture, or loss of virtue. Women hold a valued position in our society and in our military. One fifth of the Air Force is made up of women.

By comparison, Women in Israel, represent 70% of the Israeli Army Combatants. Captured Israeli women have been subject to rape, torture, mutilation and beheadings. Those types of risks go against our American values and we as a nation, choose to avoid intentionally placing women in these extreme situations.

There has also been minimal female harassment and/or molestation crimes in our military, as the increase in women service members has largely raised the visibility of such acts, and the consequences for this behavior are severe.

Enlisted Men & Women vs Commissioned Officers

If you are a college graduate, you may qualify to enlist as a commissioned Officer. If you did not attend college, you will start as an enlisted Soldier, Sailor, or Airman.

College graduates will still have to be vetted to qualify, but assuming you achieved passing grades, this is a small hurdle.

There are also programs to pay your way through medical school, and graduate as a Doctor, Dentist, Nurse, Technician, or Specialist. These programs usually require prerequisite training, and a lengthened enlistment time, to offset the training time and financial cost expended by the government on your behalf. The military also has similar programs for pilot training, engineers, & astronauts.

Enlisted service men and women have the opportunity to upgrade into these programs by first becoming a commissioned officer. After you have completed your advanced basic training as an enlisted soldier, you can take classes to overcome or meet the educational requirements to become an officer. Then you can apply to Officer Candidate School (OCS). Getting into OCS is not easy, but thousands of servicemen accomplish this every year, so don't despair, if you apply yourself, it is attainable.

In your first enlistment of three years or more, you will be focused on perfecting your skill-sets and knowledge. Typically, a three-year enlistment might result in a rank of E-3 or E-4, Which is like a "Lead" or supervisory position in civilian life.

Warrant Officers are classified as W grades. Commissioned Officer ranks are classified as O grades. W grades are typically specialized administrative or technical positions which are typically Enlisted personnel, who as Warrant Officers, have all of the same privileges as an Officer.

Rank has its privileges

The military system is based on a Meritocracy system rather than a Democratic system. By that, I mean that promotions are not subject to a vote, or some equality equation based on social biases or criteria.

Instead, promotions are based on proficiency & accomplishment. There is also a "time in grade" minimum to ensure that no one is getting pushed up the ladder ahead of anyone else who is just as proficient, but has been in their position a much longer time.

In addition, Commissioned Officers may be subject to a limit of how many Officers are needed in each rank across the service. This prevents the "Too many chiefs, and not enough Indians" syndrome that befalls many large organizations.

The military organizational chart is based upon the functional need in every position and the level of responsibility that each position requires. This ensures that Supervisors have the level of authority to direct the actions needed to be taken. Along with this level of authority, comes an equivalent level of responsibility. In the military you can delegate authority, but you cannot delegate the responsibility that goes with that authority. This reinforces that importance of when authority is delegated, it is done with proper consideration & forethought.

The Chain of Command

The Chain of Command ensures that problems are dealt with at the lowest possible level, while important issues include everyoneimpacted by the information from Private to General.

The Chain of Command works, because there is a mutual understanding and a respect for each person's rank and position. We trust in their judgement. It ensures that the priorities of the moment are resolved at the lowest level of capability and control. See the following chart.

Rank insignia	Army & Marines	Navy & Coast Guard	Air Force & Space Forces

Military Officer Ranks

0-1 Brass bar	2nd Lieutenant	Ensign	2nd Lieutenant
0-2 Silver bar	1st Lieutenant	Lieutenant J.G.	1st Lieutenant
0-3 Dbl Slvr bars	Captain	Lieutenant	Captain
0-4 Gold Oak Leaf	Major	Lt. Commander	Major
0-5 Slvr Oak Leaf	Lt. Colonel	Commander	Lt. Colonel
0-6 Slvr Eagle	Colonel	Captain	Colonel
0-7 1 star	Brigadier General	Rear Admiral 2	Brigadier General
0-8 2 stars	Major General	Rear Admiral 1	Major General
0-9 3 stars	Lt. General	Vice Admiral	Lt. General
0-10 4 stars	General	Admiral	General
0-10 5 stars	General of the Army	Fleet Admiral	General of the A.F.

Warrant Officers

W-1	Warrant Officer 1	- same designation across all services
W-2	Chief Warrant Officer 2	- same designation across all services
W-3	Chief Warrant Officer 3	- same designation across all services
W-4	Chief Warrant Officer 4	- same designation across all services
W-5	Chief Warrant Officer 5	- same designation across all services

Enlisted Ranks

E-1	Recruit	Recruit	Recruit
E-2	Private	Seaman Apprentice	Airman
E-3	Private 1st class	Seaman 1st class	Airman 1st class
E-4	Corporal	Petty Officer 3rd class	Senior Airman
E-5	Sargeant	Petty Officer 2nd class	Staff Sargeant
E-6	Staff Sargeant	Petty Officer 1st class	Tech Sargeant
E-7	Sargeant 1st class	Chief Petty Officer	Master Sargeant
E-8	Master Sargeant	Sr. Chief Petty Officer	Sr. Master Sargeant
E-9	Sargeant Major	Command Master Chief	Chief Master Sgt.

Chapter Three

Basic Training "Boot Camp"

Turning Teenagers into adults is no easy task. Turning them into adult soldiers is even tougher. Now accomplish this in only 10 weeks (Marines 13 weeks), and you are a miracle worker. That is precisely what the military does every day and has done so, for the last 100 years. This time frame is crucial for a number of reasons;

First, most teens, despite having an excess of energy are not in the best of physical shape. So, strength training is the first order of business. It starts with marching, then hiking, then running, and in between, more marching. It is crucial that these kids increase their physical stamina on a somewhat rapid schedule. This is to avoid muscle injury, boost coordination, boost the body's natural immunity, and to mentally prepare them for battlefield conditions where rapid movement and long hours are a reality. You can't make progress on the battlefield if you are dead-tired, weak, and falling asleep, so your body and psyche have to be "tuned up" so to speak, to endure the hardships of a potential battlefield situation. Physical stamina leads to mental sharpness, and the next phase of Basic Training is to develop the determination and discipline to out-smart the enemy mentally, as well as physically.

Second: Discipline is necessary in order to develop the proper instinctive behavior, as well as your proper social behavior. Military men and women are typically optimists by nature, and this is reinforced by strict disciplined training.

Knowing your left from your right seems elementary, but when you are marching in a group, you quickly see, and learn, that you have to develop your instinctive skills, and that starts with knowing forward versus backward, Port from Starboard, Left from Right, and Up from Down.

Discipline is the key to success in the military. Learning to listen, following directions, & instinctively doing the right thing at the right time for the right reason. If you find yourself in a stressful situation, your life may depend on your ability to maintain mental control. If you have been living an undisciplined life before the service, you will quickly learn that a lack of disciplined behavior has consequences in the military.

Third; Teamwork has to be learned. There is no justification for an individual to disobey orders and ignore his team-mates because he thinks he is right, and they are wrong. Teamwork involves collaboration and building trust, by working together to develop a plan, and then collectively executing the plan. Our military is successful because we have learned that when everyone works together to accomplish something, they are a mightier force than if a bunch of individuals just start doing whatever they please, and as a result, compromises the safety of everyone else. Whether it is marching as a unit, or fighting as a unit, helping and covering for each other, is the key to repeated battlefield success. Once you learn this, you never forget it.

Fourth; winning battles requires overwhelming odds. The sooner you can ramp up your battlefield forces, the quicker you can resolve a conflict. The ability to reinforce existing forces with fresh troops or additional troops, the better position you will be in to win.

Sometimes the enemy cannot reinforce their troops, and the result is a loss of motivation, or the Will-to-Win. Any military force has to be able to see a path to victory in order to be successful. When you are out-manned and out-gunned, your motivation and your momentum can deteriorate to the point that soldiers simply give up. Mental toughness is just as important as physical toughness. General Patton knew that, when during WWII, he pushed American troops to keep pressing forward, to never slow down, or stop, until victorious.

Disciplining yourself becomes second nature, and now if you have a threat on your position, or you have been awake for thirty hours on a battlefield, you will not stick your head up to look around for the enemy. You will instinctively know how to mentally keep yourself sharp and alive. American servicemen have one of the lowest battlefield fatality rates of any army, past or present. We fight harder and we fight smarter.

Basic training teaches you Tactics. How to avoid unnecessary risks, and unsuccessful tactics. Young men and women are more impulsive, and more likely to make a battlefield mistake, like walking into a minefield, or piloting a boat without consulting a navigational depth chart, or failing to pre-emptively load or arm a weapon system. If they do not understand typical battlefield tactics during the heat of battle, they will be at far greater risk.

Basic Training takes place at the following locations:

Army.......... Fort Benning - Columbus, Georgia
Fort Jackson - Columbia, South Carolina
Fort Leonard Wood - St. Robert, Missouri
Fort Sill – Lawton, Oklahoma

Navy...........N.A.S. – Great Lakes, Illinois

Marines........M.C.T.C. - Paris Island, South Carolina
M.C.T.C. – San Diego, California

Coast Guard ...C.G.T.C...Cape May, New Jersey

Air Force.......Lackland AFB - San Antonio, Texas

Space Force....Lackland AFB - San Antonio Texas

Following Boot Camp, each recruit is sent to Technical Training School. The location of these schools varies by service, and by the type of Technical Training required. There, you will be assigned to a barracks, and you will begin to settle into a normal military life.

Military Pay

Military pay in 2025 starts at $2,144.00 a month ($ 25,104.00 per year) for recruits, but following completion of Boot camp, the basic pay increases to $ 2,319.00 for achieving E-2 status. An E-4 status after three years pays $3,354.90 a month, or roughly $ 40,000 a year. Some specific job skills offer sizeable re-enlistment bonuses. Keep in mind that the military provides all your housing and food plus your salary.

In the previous chapter, we touched on the Chain of Command process. There is another reason for the Chain of Command. The Chain of Command is also a chain of responsibility. Let me illustrate this with a real-life example;

After President Reagan and Russian Premier Gorbachev had signed the Intermediate-range Nuclear Arms Agreement, in 1987, President Reagan invited the Premier, and a group of Russian Generals to tour one of our newest Aircraft Carriers. Upon concluding the tour, the Russian Generals were asked if they had any questions.

One General spoke up. He said that what had impressed the group the most, was the fact that all of above-deck aircraft moves were being orchestrated by men several decks below, who had no visibility to what was going on above them. How do you do that?

The answer was H.U.A. (Heard, Understood, and Acknowledged).

Every command that was given was repeated back. Only when it had been repeated back, did the person giving the command know that the responsibility for carrying out the command was now accepted and being carried out by the next person in the command structure.

This is the same process still used by commercial Airline pilots communicating their checklist compliance before takeoff & landing.

* * * * *

Chapter Four

Duty

What does Duty mean to you? To me, it means both Obligation and Responsibility. It is doing what is expected of you, or required of you.

As the son of a family that has a history of military service, I understood from my father, that if we want our children and grand-children to be able to live in a free country with all of the benefits that implies, we should feel obligated to not only defend our country against our adversaries, but to co-operate with those countries in peacetime, that were our allies in times of war. When countries work together, they are collectively stronger than if they stand alone. As the most advanced nation on earth, the United States is often called on to mediate disputes, and offer economic, medical, and/or military aid when requested. To that end, the United States has sacrificed both human and financial resources to achieve and maintain a world of mostly free, self-governed, nations.

Since the Revolutionary war, the United States has lost a total of approximately 1,300,000 men and women defending liberty and freedom around the world.

When the United States was still just a rebellious colony of Great Britain, it was the French who came to our aid, and ever since the revolutionary war, the French and the United States have been strong partners in defending other struggling nations, or fighting threats to our joint democracies. Today, nearly 250 years later, the French still share our desire to make the world a safer place.

When we fought in the Spanish American war in 1898, we defeated the Spanish conquest of Cuba and the Philippines. The Philippines to this day, are advocates of liberty and freedom & Spain has sent troops to fight alongside America in Afghanistan, and Iraq.

Great Britain, whom we fought during the revolutionary war, is our greatest ally. Great Britain has relied on us to come to their aid both in World War I, and World War II.

After defeating the Nazi regime, Germany has joined us, as one of our strongest allies. Japan, who was also our adversary, is another country who has become one of our strongest allies. Other European countries have also supported us, because we do not seek to rule over them in victory, but helped them rebuild as independent nations. By both our experience, and our tradition, we are not Imperialists.

You might wonder why we should feel obligated to continue to defend these global objectives, since other countries seemingly have more to gain from our aid, than we do from them. History has a way of repeating itself, and when we lose sight of our history, we either make the same mistakes over again, or we learn from them, so as not to repeat the same mistakes the next time a national or global crisis occurs. Most of these countries have experienced governments ruled by totalitarians, extremists, dictators, and religious zealots. In our case, we learned early, the value and importance of being a free country governed by its own democratic principles. As a result, we maintain a fully equipped and fully armed military to protect and defend these principles.

The Greatest Example of Duty & Purpose

When asked what was the best example of Duty, the forthcoming answer was without a doubt, the best hands-down example. It was the Landing on Normandy beach on D-day during World War Two.

That day; June 6th, 1944, involved over 150,000 Allied soldiers who were all committed to beat back the Germans, and put an end to their attempted conquest of Europe. On that day, every allied soldier, regardless of which country they represented, or their rank, was focused on this singular mission, despite the fact that everyone knew that the possibility of being injured or killed on that day or the next, was very high. They knew it was their duty, and they were prepared to fulfill it.

When the war ended in mid-1945, it was determined that about 4500 soldiers had lost their lives at Normandy. This represented about 3% of the troops that landed there. Another 5000 troops were injured, which represented another 3%. This all occurred in the first 72 hours of the initial beach landing. This has to be the greatest display of duty and purpose ever.

In the end, we were victorious. We were victorious because to a man, everyone did their duty. They fought like hell, took risks, and put their lives on the line for a cause larger than themselves.

It can be argued whether a 3% loss was worth it, or a 6% rate of out of action soldiers was worth it, but in the end, everyone knew the mission. Restore a free Europe by defeating the Nazi's and thus ending the war....... Mission accomplished!

Duty versus Religion

Our Nations' forefathers separated religion from governmental oversight in our founding documents. This was not by accident.

We had observed the extremism of countries whose religious beliefs over-rode the importance of sound economic and judicial principles.

When a religious belief becomes so overwhelming or are so encompassing, that it smothers the innovation and ingenuity of its followers, it is no longer just a religious belief, but instead evolves into a radical cult of leaders seeking control over every aspect of human life.

Clearly the cultures of the Taliban, Hamas, Islamic Jihad, and Hezbollah have demonstrated that their approach does not end suffering, poverty, nor crime. In fact, it only demonstrates that these religious beliefs actually perpetuate poverty, injustice, and criminal behavior. It also demonstrated that causing death & dying while doing it, was their religious obligation and duty. Their mantra is "Convert to Islam or die".

As time marches on, the citizenry of every country grows older and the principles that our forefathers fought and sacrificed for, start to fade from the memories of the younger generations, who never experienced the tyranny that came before them. To maintain that edge that keeps us safe from the extremists of todays' world, we maintain a peacetime Army, Navy, & Air Force.

When you choose to serve in the armed forces, you are preserving the ideals and values that our fathers, grandfathers and ancestors sacrificed for. That is why we call it our "duty." It is our obligation, to perpetuate the strongest possible military deterrence that we can. We unselfishly support any government who shares our ideals and values.

* * * * *

Chapter Five

Honor

When the news media features the President of the United States placing a wreath at the Tomb of the Unknown Soldier, do you really think about why we honor someone who is "known but to God"?

The Memorial Tomb represents the sacrifice that all the young men made, who died while serving in the military. It was dedicated by President Warren G. Harding on Armistice Day (November 11th) in 1921, following the end of World War I. It contains the remains of three unknown service members who died in that war.

For the most part, they were young men, many of whom were just teenagers, who had their bodies blown to bits. Their body parts unidentifiable. Many were buried on the spot where they were killed. Their loved ones never knowing what happened to them or even when or where their remains are. At most they got a telegram. If they were lucky, they got a letter from a fellow soldier who served with their child. Some even got a visit from a military attaché.

But for the hundreds who never got to bury their loved one, something meaningful had to be done. The result was a memorial to honor all of them, in perpetuity, lest we never forget. As the military post-script often states; "*All gave some, and some gave all*".

Given that somber opening, that is what Honor is all about. It is a quiet demonstration of respect. It is an acknowledgement that the cause was worthy, and in recognition of the sacrifice every soldier made pursuing it. They did their duty. That is what honor is all about.

So, when you choose to put on a uniform of the United States Armed Forces, you are demonstrating your own acceptance of the responsibility and the obligation of every soldier who came before you, and those who will train alongside you, and then it is <u>your</u> Duty to set the example for those that will someday come behind you.

If for no other reason, you can take pride in your uniform, as one of the most outward displays of Duty, Honor, and Purpose.

Pride

You don't hear the word Pride used much in society, but when it comes to our military, you hear it being used to describe the one aspect of our country that people actually do take pride in.

In addition, our servicemen are proud to wear a uniform that represents the power and strength of our government, and it makes them look and feel respected, dignified, and valued. If polled, most of America would tell you that

- We are proud of those who serve in our military
- We are proud of the capabilities of our military

- We are proud of our military's contributions to the world
- We are proud of the sacrifices of our military
- We are proud of the positive image of our military
- We are proud to honor our fallen

There are many honorable professions; Doctors, Nurses, Police, Firemen, and yes, even Lawyers (sic). These professions are considered honorable because they require specialized and intensive technical skills, a high degree of intellect & integrity, and coupled with a long-term commitment to help our fellow citizens every day of their respective lives.

We call them honorable because we appreciate their efforts to assist us through difficult situations. We call them first responders, but they are more than just the first person to come to our aid, they possess a degree of training that is not shared by others. They care enough that they are prepared to risk their lives and/or their reputation to save yours. As a result, we call these professions; honorable. These are individuals, who by their title, skill, and demeaner, we respect. The members of our military fall into this category as well.

Being a current or former military freedom defender requires that you understand and appreciate the sanctity of life, or that you are prepared to defend against someone else's involuntary death, even if it means sacrificing your own. The fact that you are prepared to so, is an honorable act of the highest Order. Your honor is your promise.

Necessary & Unnecessary Risk

Conflicts of any kind are a messy proposition. One thing is certain, freedom is not free and those who are unwilling to fight to hold onto it, are destined to lose it.

One of the important lessons to learn in life, is what is a necessary risk, verses an un-necessary risk. The difference can mean the difference between being a survivor versus a statistic.

Learning to evaluate risks is a skill that translates into success on the job in private industry, just as it does in the military. In civilian life, a risk may mean a financial risk, an operational risk, or a tactical or strategic risk. Military tactics are no different.

"*Evaluate, then Execute*", whether it is a military operation or a civilian operation. Being decisive in these situations is often the difference between success and failure. When you delay making an important Mission decision, you are losing time and gradually losing your best options. Time can often be your best weapon.

If you are successful, you will be honored, if you are not, you will still be remembered, and you will be honored for your effort. Honor is the recognition of the commitment you made to faithfully fulfill your mission to protect our nation and our citizens from adversary's. A notable motto is "Die for a cause, not 'cause you didn't care".

Statistically, if you are in the U.S. Army, you have the greatest chance of being injured or killed. Those odds however, are very low. Lower than your chances of being killed in an accident in civilian life.

The Marines, despite usually being the first military group thrust into a conflict, have slightly better statistics, where the likelihood of death and injury is actually lower than in the Army. The Navy, Air Force, and Coast Guard have even lower death or injury rates as most of their action is taking place far from the battlefield. None the less, every combat role poses a risk. So, what is that risk?

According to the Army, your chance of being killed in action in the Army, is about ½ of 1 %. The reason for this low percentage is quite simply the fact that 90% of the positions within the Army are in support functions, not direct combat. Add to that, modern weapon technology makes direct man-to-man combat far less common than the wars of the past. Medical breakthroughs over the years, have also saved many lives that in the past might have been lost.

If you are worried about being pre-maturely killed in action, my first advice is to be sure and look both ways before crossing the street when you leave home. Your chances of being killed or injured by an automobile in a civilian setting is much higher.

Death & Disability

If you want to know what "Honor" is all about, visit a few of the finest memorials to servicemen that you will ever encounter. The "Tomb of the Unknown Soldiers" in Arlington, Virginia is a good start, or while you are in the D.C, area, the Viet Nam war memorial is yet another, or the Korean war memorial, or the WW II statue of our flag being raised victoriously over Iwo Jima.

If you vacation in Hawaii, visit the war memorial in Pearl Harbor, or travel up to Punchbowl cemetery. Closer to home, visit any Veteran's cemetery anywhere in Europe or the United States. At each of these locations you will get a sense of what Honor is all about. These are some of the most stunning and well-kept memorials you will ever experience. All of these places exist for one reason; to honor those who sacrificed their lives in the defense of your freedom, and others around the world.

We honor them, because it is the very least, we can do to recognize their contribution and their sacrifice, to ensure that the rest of us get to live a life free of fear, subjugation, or harm. These are the things we take for granted, but they only survive because of the sacrifice of others. To honor them, is to express our appreciation in a permanent and meaningful way that says to the world, we value our freedoms, & those who defended it

One may ask why a soldier throws himself on a live grenade, rather than try to run. The answer is this: He knows he cannot outrun the deadly explosion that will likely happen in the blink of an eye. His fate is already sealed. By throwing himself on the live grenade, he shields and ultimately saves the lives of everyone else around him. It is not only the honorable thing to do, but it is the only action he can take that will enable his fellow soldiers to keep fighting for what he was fighting for.

The Arizona Memorial

While most Americans have seen the Tomb of the Unknown Soldier on television, many have not seen one of the most beautiful war memorials ever constructed. Located in Pearl Harbor, Hawaii, bridging the sunken battleship; U.S.S. Arizona, is a white structure that pays tribute to the 900 men trapped below, when the ship exploded during the Japanese attack on naval forces stationed there.

Designed by Alfred Preis, and paid for with mostly private funds, this memorial can only be reached by a special tour boat operated by the Navy, which transports tourists and visitors to the memorial. The names of those lost are posted on a white marble wall while opposing stained-glass side windows bathe the names in a beautiful glow of shimmering colors, as the water below reflects the sunlight through the rainbow-colored glass.

It is the manner in which the military honors its' fallen that is most impressive. Authorized by President Eisenhower, the memorial was to be built from private funds.

Elvis Presley, himself a former Army soldier, helped raise money for the floating memorial, and personally donated a very large sum. He also raised over $60,000.00 from performances to help build the memorial, adding it to the over $200,000 raised by the American Veterans (AMVET) organization.

If you are ever in Honolulu, don't miss it.

* * * * *

Chapter Six

Purpose

If someone asked you what the purpose of your life was, or to put it another way, when your life is over, and you are standing before the gates of heaven, and you are asked what you did with your life, there is really only one acceptable answer; "You left your world in a better place than when you arrived there".

By "*your world*", I mean that corner of the globe that you had control of. In many cases, this is your home, your school, your workplace, and your community.

It does not matter whether you lived for thirty years, or lived for a hundred years. What matters is, what you did with the gift of life that you were given. What did you do to make your world a better place? Would your family speak of you with pride, or disdain? How did you contribute to a better society?

When you are a first responder, whether it is as a civilian, or as a military defender, you learn that honorable professions provide you something that most people strive for, and struggle to achieve and rarely find; Purpose……..Some never find it.

When your life has purpose, you can focus on enhancing and honing your professional skills, your knowledge, your experience, and become an expert in your field, all while making your world a better, or safer place, for the rest of us.

Living a life of purpose develops your confidence and leadership skills, as well as sharpen your physical and mental capabilities. When you help others, you feel good, just as you made them feel good.

When we refer to a purpose-driven life, we are talking about a direction. Working towards something. As the old adage says;

"If you don't know where you're going, any road will get you there".

Too many people today are living life without a purpose. It is not enough to simply get married, raise children, and then slave away at a job you don't like, for a boss that you don't like, working for a company you don't like, etc. etc. Don't live a miserable life. Be happy!

It has been shown that people who lead a purposeful life, live longer

The key to a purposeful life is to have direction and then move in that direction to accomplish or contribute to the effort. It is to believe in yourself, and to believe in others.

Some people are great volunteers. They offer to help where they can, with what they have. Sometimes. It is just a caring word, a glass of water, or a hand up. Almost any church volunteer can help you.

Then there are those I call "the Slackers". These are the sponges of the world. They take whatever they can get, while doing as little as possible. We see them sitting in entryways, sitting on the curb, or maybe demonstrating about why they want even more free services & sympathy from the rest of us. If you truly need help, visit a church.

When I see a seemingly healthy 25-year-old adult standing on a corner, looking sad, begging, I wonder, why don't you join the military? After all, you will get nutritious food, a warm place to sleep, a job, and a regular paycheck, and most importantly, a purpose in life. Anyone can be down on their luck. It's those that pick themselves up and establish a direction that get my sympathy and assistance.

When you are in the military, you understand that the particular service you are in has a mission. They have a Peace-time mission and a War-time mission.

In Peace-time, the mission usually revolves around the Safety, the Security, and Operational Readiness of our military manpower, our War-time machinery, our supply line of materiel, and the care of our various war time munitions.

Every aspect of the military trains, and re-trains. All that repetitive training builds proficiency. When conflicts break out, the men and women of the armed Services have to be fully prepared and fully equipped to carry out the missions that resolving the conflict requires in a moment's notice. Proficiency reinforces Efficiency.

During war-time, the organization morphs into one of action. Millions of pounds of supplies have to be airlifted, or shipped. Thousands of servicemen have to be transported. Tanks, trucks, weapons, ammunition, food, and water may have to be moved halfway around the globe in 24 hours or less, depending on the scope and size of the threat. Aircraft may have to have a safe air base to operate from, as well as shelters for Command communications and satellite equipment.

All of this logistics know-how provides great career opportunities in civilian companies when your service enlistment is over. Jobs in Distribution Management, Transportation Management, and Network Management are all highly paid positions in private organizations.

Patriotism

If you are a Patriot, you are in essence a "cheerleader" supporting the principles and values of your country and your leaders. You take pride in your country and are respectful of its government.

All too often, individuals selfishly put their own principles ahead of those of the country, believing that they have a right to impose their values on everyone else. That is not only anti-patriotic, but borders on anarchy, which is the opposition to any authoritative hierarchy, and advocates abolition of a centralized government.

Anarchy usually reveals itself as rebellious behavior against any governmental authority and typically escalates into the defacing or destruction of public (government) property. Anarchy offers no alternative, but simply the elimination of all government authority.

Extremists, on the other hand, typically promote an alternative form of government, but these alternatives usually dictate policies that are illogical, immoral, or reckless, and whose aim is to simply swap one form of community standards for their own.

In its mildest form, it is civil disobedience, and at its worse, it represents acts of treason.

* * * * *

Chapter Seven

Respect

Why do military people salute? The Military salute is a show of respect. It is respect for the person's rank, which is a visual acknowledgement of the level of authority and responsibility consistent with that rank, and it is respect for the man or woman wearing the uniform who have earned the right to display that rank on their uniform. It is acknowledgement of their accomplishments.

Civilians do not salute, but the respect for the uniform and what it represents are understood by all. It is an unspoken acknowledgement of respect. A soldier's failure to salute an officer is viewed as an act of disrespect, and can be a punishable offense in the military.

The military uniform represents all of those who sacrificed for our freedom in the past, and who are willing to sacrifice for our freedom in the future. It represents a capability of the person wearing it, as well as a commitment of loyalty to our flag and the country it represents.

Respect is not given out like a souvenir T-shirt at a rock concert. It is earned. Earned by the effort being made every day to keep us safe, and earned by demonstrating that the wearer has what is takes mentally, physically and morally to be a competent Soldier, Sailor or Airman. You won't trust someone you do not respect.

When you join the Amed Services, you are expressing your loyalty, pride, and commitment. Your medals & ribbons are proof of your courage, your loyalty, & your accomplishments.

Many of the ceremonial rituals of military service demonstrate this characteristic of mutual respect openly, when the Commanding officer takes over or retires, when the deceased are handled and processed, or when you watch the ceremonial guards at the tomb of the unknown soldier.

Resolving Conflicts thru Respect

Conflicts usually develop because there is a lack of mutual understanding or agreement about the facts, or the circumstances. Both parties, or neither party, cannot reach acceptance or agreement, without first having a common understanding of the circumstances. That in itself, is sometimes difficult. It usually won't happen unless there is respect for the other parties' position. Once there is respect for the other sides' position, a mutual discussion and/or agreement can begin to develop, which in turn builds trust. Without respect, there can be no trust.

Here is a real-life example;

In 1968, North Korea attacked and seized the USS Pueblo.

The United States maintained that the Naval Intelligence ship was in international waters, but the North Koreans disagreed, and held the ship and its crew for 11 months.

President Lyndon Johnson refused to apologize for a mistake that he claimed was clearly North Korea's, and the two sides were at a stalemate.

When Richard Nixon was elected President in November of 1968, Dean Rusk, President Johnson's Secretary of State, convinced Johnson to resolve the issue before leaving office in January.

The issue was that the North Koreans wanted a formal apology, and a signed confession about spy activity by the ship.

It was finally resolved when the U.S. side came to understand that in North Korea and other Asian cultures, an apology is not an admission of guilt, as we often think of it here, but rather a simple show of respect.

In addition, North Korea had a pattern of admitting to things in private, but disavowing them in public. Privately, the North Koreans admitted that they might have made a mistake, but publicly they insisted that the U.S. had made the mistake, disavowing their private acknowledgements.

The issue was resolved when President Johnson publicly apologized, and then the U.S. signed a private confession, which they immediately publicly disavowed, the same as the North Koreans do. This display of respect was ultimately what led to the resolution of the conflict, although the ship is still held by the North Koreans to this day (57 years later).

This story is a great example of how different cultures can talk past each other because they simply do not understand the culture of the opposition. Language differences complicate discussions, but the nuances of language are equally important.

Mutual Understanding leads to Respect for conflicting points of view. Once two sides can respect each other, they can create mutual Trust. Here is a way to think about it;

- Understanding leads to Respect. Respect leads to Trust.
- Trust leads to Loyalty. Loyalty leads to Duty.
- Duty leads to Purpose. Purpose leads to Courage.
- Courage leads to Honor.

During the Carter administration, Israel and Egypt were not on speaking terms. In fact, they despised each other and did not trust each other. Israel had defeated the Egyptians in the so-called six-day war and President Jimmy Carter, having talked to both sides at some length decided that the only possible way to resolve their differences was for the two sides to meet and talk face-to-face to one another.

After many back-channel discussions on minor issues, President Carter invited Menachem Behgan, the prime minister of Israel and Anwar Saddat, president of Egypt, to come to Washington to hammer out the bigger issues that neither side was willing to address.

Carter insisted that they spend their first visit at Camp David and that they focus on their personal lives rather than their respective political disagreements. He wanted them to get to know each other on a personal level, as he had done, because he felt that in his private conversations with each man, there were a lot more common-ground interests, than there were differences. He knew that there were geographical, economic, and commercial issues that could potentially unite the two, but the first step was there had to be mutual respect between the two, if any progress was going to be made. After several days, the two sides started drafting the framework of an agreement that addressed many of the concerns each side had, while focusing on issues where cooperation was of mutual benefit. There were two agreements, each covering different aspects of their differences.

After each side went home and discussed the details with their respective lawmakers & military leaders, the two sides communicated to each other directly that an agreement was possible. They agreed to go back to Washington, openly discuss their concerns with Carter, and potentially sign an agreement, if Carter would agree to stand behind their guarantees to each other. Carter agreed.

The Camp David accords were signed in a white house ceremony on September 17th, 1978.

The cooperation and agreements between the two countries have largely been adhered to, and there have been no military conflicts between the two countries for the last 47 years.

In 2020, the Abraham Accords were signed by several Arab states and Israel in a similar scenario which was instigated by President Trump. Like Carter, he knew that there were mutual benefits to be had by these countries by cooperating with Israel and acknowledging its right to exist. The continued animosity over issues was that they were more about a Palestinian homeland, than they were over the economic well-being of their own populations.

Bahrain, Morrocco, the United Arab Emirates, and Sudan all signed on. Only Sudan has struggled to maintain relations with Israel, due to its large Palestinian population, which vehemently opposed the agreement.

These regional agreements have global implications, but they all started with reaching a common understanding and subsequent respect for the views of their adversaries.

* * * * *

Chapter Eight

Trust & Faith

How do you build trust in an army of thousands of soldiers or sailors who barely know each other? It starts with a common understanding of what they are fighting for.

Members of the Armed Forces share common values, common desires, and a common mission. They have learned that they can depend on each other, knowing that each service member has been exposed to the same tactics, and has completed the same training. Learning to trust the person next to you, and to trust that you can rely on them to watch your back, is crucial on the battlefield.

This teamwork of depending and relying on others, knowing they are relying and depending on you, builds teamwork, which in turn reinforces trust in the capabilities and confidence of the person you need to rely on., which creates faith in the system. All the services share a common understanding that if you are killed or injured, you will not be left on the battlefield to die alone. "Leave no man behind" is more than a slogan, it is the philosophy that everyone is a "brother" and deserving of the honor, care, and respect that you would provide your own brother. We are "Brothers in Arms", affirms that we have faith in that military commitment.

Loyalty is viewed as a personal commitment to act, while Trust is the belief that you will act. It is the reliance or faith we have in others, to do what is needed, when it is needed, and where it is needed, without questioning why it is needed. When people do act and fulfill out trust in them, we respect their commitment.

Trust also creates integrity. Integrity is practicing open and honest communications, and having compassion. When you observe the ceremonial rituals of miliary service, albeit an officer's retirement or watching the ceremonial guards at the Tomb of the Unknown Soldier, you are witnessing rituals that demonstrate true integrity.

Cowardice in the military is not just a refusal to fight, it is a loss of trust that you will actively protect others just as deliberately, as they protect you. Some individuals claim that they are a Conscientious Objector and that they cannot knowingly, or willingly kill another human being, regardless of what side they are fighting for. This is seen as a lack of trust, in that you can't be depended upon to protect a fellow soldier. The enemy does not share your beliefs. They are out to kill you, & everyone associated with you. You are only an objector till they come for you.

During armed conflicts, you are depending on, and trusting, that each of your fellow soldiers are protecting you and/or your position from harm, just as you are protecting theirs. The military does not pass judgement on a soldiers' beliefs, but they know that a lack of trust is a sure recipe for disaster under fire. Survival is often the result of your faith in the system and its commitment.

Decisive action is a military behavior that has to become second nature. The slightest hesitancy to act can mean the unnecessary loss of life, so Trust is a key attribute taught in the military and cowardice is not tolerated.

Inspiration

If there is one thing you will learn to appreciate in the military, it is the power of inspiration.

When you work for inspirational people, you are motivated and you learn to trust them without hesitation. You admire them. Inspirational leaders are not only leaders because they have a record of successes. It is usually because they empower their subordinates to take charge of their tasks, and do what's right, and/or do what's best. They demonstrate their appreciation. Your ribbons/medals prove it.

That empowerment is a given level of trust in you as a person, and in your ability. You are viewed as a person who is educated (knowledgeable), trained (skilled), and capable (motivated). True Leadership is confidence in your ability to accomplish the mission. This is the highest level of trust a leader can instill in his subordinates. It is recognition & acknowledgement of your abilities.

Success accomplishes three things; It reinforces your belief in yourself, it builds trust among your subordinates, and it reinforces the trust in your leadership.

As your confidence increases, you tend to look at situations as no longer *impossible*, but rather simply a challenge to determine the best way to *make it possible*. Inspiration often leads to innovation.

Trust becomes a combination of mutual admiration, faith, appreciation, recognition, confidence, inspiration, & innovation.

Having Faith

Faith is a core tenant of military service. See the following story:

A Chilling Story of Faith

During the fall of Saigon near the end of the Viet Nam war, a personal friend of mine; "Jim", was a Marine guard charged with defending the U.S. Embassy. On that fateful day, the order was given to helicopter out any Viet Nam-ese civilians. The Viet Cong were now over-running the city and would soon be attacking the barricaded embassy. Everyone in the embassy fled to the roof.

When the last helicopter was fully loaded, the pilot yelled out that this was the last ride out, and the guards would have to remove five locals to make room for the guards. The guards looked at each other and the desperate faces of those on the helicopter, and waved the helicopter off. "*Just tell God we're here*", a guard yelled.

They would defend the embassy to the last man. Anyone left behind would likely be killed by the Viet Cong. The remaining civilians learning the air-lift was over, fled to safety.

After about 15 minutes, an empty helicopter appeared and the pilot said "*This is it, last bird out!*". No one had expected another helicopter, so the guards all piled in, and they were flown to safety.

Upon arrival on a Navy ship, Jim asked who the pilot was, as they wanted to thank him. The deck officer said, he did not know, as the helicopter had already left the ship. They called the only other ship with helicopters, and they said it was not one of theirs. It must have been an Army helicopter, not a Navy one. The Air controller said no other helicopters were assigned to their area. Who saved Jim and his fellow guards? ....God only knows.......... Thanks, God!

* * * * *

Chapter Nine

Loyalty

When we pledge allegiance to the flag of the United States, we are pledging our loyalty to our flag and the country it represents, and its citizens. Loyalty is not like a coat that you put on and take off as the circumstances change, Loyalty is the commitment that we make, to be responsible for, and uphold the principles and values our country represents. We call these attributes; Freedoms.

Some idealists seem to believe that they can choose to be loyal when it suits them, but that is not loyalty, that is called dis-honesty, mis-trust, or at its worst; treason.

Loyalty to the flag is called "Patriotism". Loyalty is learning, believing, practicing and supporting a set of life-guiding principles that are unwavering. There is no grey area, either you are loyal, or you are disloyal. Loyalty is not a habit that you can kick when you are tired of it. Loyalty is the very fabric of values that people live and die for. It's knowing what is right and being willing to fight for it. It is why the revolutionary soldiers were called "Patriots". Loyalty is a personal commitment, a promise, that despite conflicts, or adversity, you will be loyal to the original principals you accepted, whether you joined the military or not.

You are accepting responsibility and stating your commitment to remain responsible. When we read about traitors, we see the consequences when people compromise their responsibility and their loyalty to the basic principles of our democracy, including fairness, honesty, and security.

I was once challenged to explain to my peers, what loyalty looks like in real everyday life. Here was my answer.

"If your car breaks down at 3 AM, thirty miles from home, and your parents or siblings are away on vacation, you call your most loyal friend and wake them up, and ask them to come get you. Your friend does not remind you what time it is, nor suggest you call someone else, they simply ask where you are, and apologize that it will take a while to get there."

That to me, is loyalty. A level of understanding respect, and trust, that is bundled together in friendship. It is knowing that if this event happened to them and they called you, you would do the same thing. It is an absolute, unwavering commitment to do the right thing at the right time for the right reason. No questions, no hesitation, no doubt. You cover me, and I will cover you. It's trust at the highest level.

Once in a while, people confuse loyalty with courage, but there is a difference. Loyalty may involve inconvenience, or even reluctance to act. Courage on the other hand, is acting in the face of danger, because there is no one else that can act, or will act, so you do.

Loyalty, true loyalty, is more than just the reciting of the Pledge of Allegiance. It is a personal promise to uphold and fight for the values, principles and freedom that our country was built on, and stands for. It is our promise to our God, to our family, our friends, and our community that we will do our best to defend our freedom against those that would compromise or eliminate them.

Loyalty is the ultimate expression of trust.

Loyalty Under Fire

At the beginning of this book, I mentioned that I was one of five teenage boys that joined the military. My good friend Carl, had joined the Marines, and shared this personal Viet Nam experience with me.

It was a routine day, until a call came in, that a squad of Army soldiers were under attack a few miles from our position, and needed men and firepower, Carl's squad was to go in and help the soldiers. After several miles of jungle, they could hear the gunfire and every one was now on high alert for snipers, land mines, and booby traps.

When they managed to shoot their way into a clearing, the found only about two dozen soldiers were able to fight back, most being dead or severely wounded. The Army had come upon a battalion of Viet Cong and were badly out-numbered and out-gunned. The Marines, while a huge morale boost to the soldiers, were hard pressed to beat back the enemy, and soon they too were taking hits, as the enemy had effectively encircled the combined U.S. troops.

Carl worked his way to the edge of the clearing and found a tree just wide enough to allow him to stand and return effective fire. An Army soldier seeing Carl, ran over, and tried to stand behind Carl, so he too could fire back effectively. Carl did not know the young soldier, but he warned him that the two of them together made too big a target, and that he should pick a different tree so that they could multiply their firepower and be better protected.

Suddenly Carl felt the soldier slide off his back and hit the ground. Carl did not dare turn to see how badly he was injured, lest he would be exposing himself to the same shooter. He tried to reassure the young soldier, but no response. Carl just kept shooting.

Fifteen minutes later, the Air Calvary arrived in several helicopters and the enemy stopped firing and retreated back into the jungle. Carl struggled to pick up the limp soldier, slung him over his shoulder and tried to run in a hunkered down position. He barely made it to the helicopter. A soldier grabbed the soldier and pulled him on-board, then helped Carl get in, as well, Several more piled in and they all safely returned to their base camp.

The Medic informed Carl that the soldier he had carried was dead, shot through the head, probably died instantly. The Medic thanked Carl for bringing the boys' body back, despite great risk to himself, and said he was sorry it was all in vain.

Carl responded that he thought if the situation had been reversed, he hoped that the young man would have done the same for him. "Leave no man behind, that's the commitment we make to each other". [All military members take a Loyalty Oath when they enlist].

Ironically, Carl never looked at his momentary friend's face. "*I could not tell you what color he was, whether he was dead or alive, or what rank he was. I just knew that I had an obligation to not leave him there in the jungle to die by himself. He needed help, and I guess I was the guy God sent to help him*".

This is the best example I can give you regarding the mentality of those who are trained to protect us. Even under the most horrific circumstances, there is a loyalty to each other that is unspoken.

Lee Marvin, award-winning actor and war hero, was himself a former Marine. He never talked about his military heroics, but on his death bed, his last words were; "*Once a Marine, always a Marine!*"

That is Loyalty! R.I.P. Lee.

* * * * *

Chapter Ten

United States Army

"Esto omnia quae esse potes" - Be all you can be

The United States Army was established in June of 1775, following the success of the Continental Army which was formed from the local colonist Militia's to fight the British.

Our Army has largely been defensive in nature. We are not out aggressively trying to conquer other nations. Our military focus is defending our home territories. To that end, we equip our military with the best weapons and technology, and train our troops to use the best strategies and tactics in order to be successful in any conflict.

That being said, like any Army, our military does train and maintain an Offensive team that is prepared to take the fight to the enemy, instead of letting them take it to us. From the "Green Beret's" to the "Big Red One", the U.S. Army has the reputation of prevailing in every conflict with both mental and physical fitness being the primary attributes of every soldier.

Army Mission Statement

"Deploy. Fight, and Win our Nation's wars"

You will note there is no reference to social themes, financial goals, or political objectives. If there is a conflict that threatens our way of life, we will take up arms to fight it, and we are fighting to win, not just to reach a political stalemate. The motto of the Army since 1774 has been "This, we'll defend"

How the U.S. Army is Organized

Like any large organization, the Army is broken down into manage-able units which can be combined into larger units depending on the need and the circumstances. The Army uses a Land-Sea-Air triad of forces.

During war, where the circumstances require the movement of large numbers of men, materials, and equipment, various units are put under a single command. Where there is a need for specialized equipment, and men with specific technical skills are required, smaller units may be formed.

Currently, the U.S. Army is made up of 29 Divisions. These divisions include 15 Infantry divisions, 9 Training divisions, 3 Airborne divisions, and 2 Armored divisions. Each division encompasses from 12,500 to 15,000 men, resulting in a peacetime army of roughly 365,000 soldiers.

These 29 Divisions are typically divided into five Corps totaling from 75,000 to 90,000 men each, which can then be combined into 5 Field Armies totaling 450,000 men each.

During military conflicts, the President can authorize additional Corps and Field Armies*.

Each Division is broken down further into specialized Regiments, then Brigades, then Battalions, then Companies, then Platoons, and at the lowest level a Squad of 6-10 soldiers.

Commanding these groups are four-star Generals over the Field Armies*, Three-star Generals over the Corps, and Two-star Generals over each division. Lower ranked officers manage the smaller unit break-outs. * 1 Field Army consists of 5 Corps/5 Corps consists of 9 Division's.

Distinguished Army Units

The **First Armored Division**, the most well-known Division in recent history, the "**Big Red One**". Arguably, the most celebrated division in World War II, and the first division to be deployed to Viet Nam. First Division soldiers have a red numeral 1 on their uniform.

The **Second Armored Division**, who were victorious in the Battle of the Bulge Italy and southern Europe during WW II, were legendary in their courage, their persistence, and their bravery.

The **Third Armored Division,** commanded by General George S. Patton, and known as "Patton's Own". Instrumental in the battle to gain control of Italy, Southern Europe and Northern France

Distinguished Army Generals

Honorary 6 star (******) General of the Army 5 star (*****)
Supreme Commander (****) Lieutenant General 3 star (***)
Major General 2 star (**) Brigadier General 1 star (*)

John J. "Black Jack" Pershing; [1860-1948] Is probably the most influential General of all times. He served with the 6th Calvary in the American Indian Wars in 1886, commanded the Buffalo soldiers in the Spanish-American war in 1898. He fought in the Philippines, and charged up San Juan Hill in Cuba.

He fought Pancho Villa in the Mexican revolution of 1916, and then led the U.S. effort in Europe during World War I, and was named 5-star (5 *****) General of the Armies in 1919.

He was the attaché to Japan, then Chief of Staff at the White House during President Wilson's 2nd term. He taught at West Point & mentored George S, Patton. He won a Pulitzer Prize for his book on WW I. **He died in 1948, & posthumously awarded, the only 6-star General of the Armies in history.** He died at age 88.

George S. Patton; [1885-1945] Immortalized in the Movie "Patton", **Patton was probably the most successful and most arrogant figure in the history of the Army.** Patton was a Battalion leader in WW I under General Pershing in the Mexican Revolution, led the 2nd Armored Division (tanks) in WW I, and then led the 7TH Army in the battle of Meuse-Argonne, where he was wounded. He received the Distinguished Service Cross. Nicknamed "**Old blood & guts**"

He was then put in charge of the 3rd Army "Patton's Own", which commanded northern France in support of the Normandy invasion. His final rank was that of a **4-star General.** He died at age 60 in a car accident in Europe in December of 1945.

Douglas MacArthur; [1880-1964], **4 star **** General.** Graduated first in his class from West Point in 1903. He was in the Philippines during the Spanish-American war, and then commanded the 42nd "Rainbow" Division, and rose to the rank of Brigadier General during WW I. He then developed the Defensive strategy for the Philippine government in 1935, and retired in 1937 at age 57. He was awarded 7 Silver Stars.

In 1941, he was recalled to active duty, and became the **Supreme Commander of Allied Forces in the Pacific,** during WW II. He oversaw the occupation forces in Japan, & was the U.N. Commander in Korea in 1950. He died in 1964 at age 84.

Dwight D. Eisenhower; [1890-1969] **5 star ***** General.** Graduated from West Point in 1915 and assigned to the 19th Infantry Regiment. He was in a Tank Training battalion stateside during WW I, where he earned the Distinguished Service Medal. After WW I, he became an aide to General Douglas MacArthur.

In 1942, he was appointed Deputy Chief of War Plans, and in 1944, he became the **Supreme Commander of Allied Forces in Europe** (4 ****). He directed the assault at Normandy, led the Allies to Victory in WW II. In 1952, he was elected our 34th President. In 1961, He was named General of. the Army. 5 *****. Died at age 78.

John Thomas Corley (1914-1977) **Brigadier General - 1 star. One of the most decorated soldiers in history.** Graduated from West Point. Started as a Platoon leader, and worked himself up to Lt. Colonel in just three years. Numerous hand-to-hand combat victories. Considered an inspirational leader and a fearless fighter.

Awarded Congressional Medal of Honor, Distinguished Service Cross, 8 Silver Stars, Medal of Valor, National Defense Medal, purple heart, Ranger Hall of Fame. Served in WW II & Korea. Died at age 62.

H. Norman Schwarzkopf; [1934-2012] **4 star (****) General**. He graduated from West Point in 1956. A highly decorated veteran of Viet Nam, he commanded the 24th mechanized division, then participated in the 1983 Grenada invasion, then successfully ended Iraqi's invasion of Kuwait in 1990 in just 100 hours. **He was the only person to receive a Distinguished Service Medal from the Army, the Navy, the Air Force, and the Coast Guard** and was Knighted by Queen Elizabeth. His nickname was "**Stormin' Norman**". He retired in 1991, and died in 2012 at age 78.

John "Jack" M. Keane [1943-Living] **4 star (****) General. One of the most influential Generals of the 20th century.** Born in New York, Jack graduated from Fordham University, U.S. Army War College, Command & General Staff College, Gabelli Business School, and W. Kentucky University. Honorary Ph.D.

He saved the life of Gen D. Petraeus, who was shot during a live-fire exercise & would have bled to death if not for Keane's actions.

He is an Army Ranger, a Master Paratrooper, Combat Air Assault, & served in Viet Nam, Somalia, Haiti, Bosnia, Kosovo, and the Middle East. He has been awarded a Silver Star, Bronze star, Legion of Merit (5), Army Commendation medal, Army Distinguished Service Medal (2) plus over 20 more. He turned down a Secretary of Defense nomination twice. He was given the Presidential Medal of Freedom.

Tommy Ray Franks; (1945-Living) **4 star (****) General**
Born and raised in Oklahoma, Tommy dropped out of the University of Texas and joined the Army as a Private. Through the Army's continuing education program, he resumed his University of Texas curriculum and, then graduated from the

Armed Forces Staff College and the Army War College. He was called the "**Soldiers General**".

He was assigned to command the 1st Calvary Artillery Division in South Korea, and then commanded the 2nd Infantry Division, then the 3rd Army Division. He was named Assistant Division Commander of Operation Desert Shield and Operation Desert Storm. His troops captured Saddam Hussein. He was awarded the Defense Distinguished Service Medal, Army Commendation Medals (2), Legion of Merit, Bronze Star, plus many others.

* * * * *

Chapter Eleven

United States Navy

"Semper Fortis" - Always Courageous.

The U.S. Navy was founded in October 1775, in response to the Revolutionary War, and was officially established as an arm of our Defense department in 1798. At that time, the Marines were simply a sea-going fighting force of the Navy that could deter adversaries on land or at sea.

Over time, it evolved into an administrative organization that designed, built, and supported the government's sea-based ships, and developed an operational command structure to actually staff, operate, and maintain the fleet while at sea.

The Navy is the core of our nations' Defense Triad. It manages the Nuclear Submarine fleet, The Aircraft Carrier led sea-based Aviation Strike Force, and a global ocean-going fleet of Offensive and Defensive war ships. It transports supplies & ammunition for the troops overseas. It also supports NASA and the U.S. Space Force.

The Navy Mission

"To train, equip, and maintain combat-ready Naval Forces capable of winning wars, deterring aggression, and maintaining security and freedom of the seas".

How the U.S. Navy is Organized

The Navy is organized differently than the land-based Army. Instead of "Field Army's", the Navy commands "Fleets". The Navy has Three high-level command structures; One for Shore-based Naval Operations. And one for Sea-based Naval Operations.

A third Marine Corp Operational Command is commanded by the Marine Corp Commandant. A modified command structure for each Command is used during war time. Combined, they constitute a Navy Command that is commanded by the Chief of Naval Operations.

The Administrative Command encompasses shore-based Naval activities, which focuses on Support Personnel; Training, Intelligence, and Cyber warfare, plus Fleet deployment, manpower, fleet maintenance & logistical supply support.

The Operations Command is focused on managing the various sea-based Naval Fleet components, (ships, submarines, aircraft, weapons and technology).

Structurally, at the lowest level, the Navy consists of Elements. Multiple Elements equal = Units. Multiple Units equal = Groups. Multiple Groups equal = Fleets. Multiple Fleets equal a Task Force.

During war times, there would be a Fleet Admiral, who would be the equivalent to a 4-star General. In the past, there have only ever been four such "Fleet Admirals". Currently, in peacetime, there is no one holding this rank. fleets are now typically led by a Vice Admiral.

There are 7 Fleets, each led by a Fleet/Vice Admiral, who is equivalent to a three-star General in the Army. Fleet size can vary depending on the number & type of ships or aircraft in each Group. The Fleet Admiral commands "**Task Groups**", led by Rear Admiral's, who are equivalent to one-star generals. There are 11 known Task Groups.

Operating under the Task Groups are three types of "**Task Forces**". A Submarine Task Force, an Aviation Task Force, and/or an Expeditionary Task Force. Task Forces are made up of **Task Units**. The exact number of specific Task Units within each Task Force or Group is currently Classified.

"Task Units," are made up of several **"Task Elements"** each Task Unit is commanded by a Lt. Commander [O-4] depending on the nature of the Task Unit's mission. It is estimated that there are about 200 Task Units, based upon the Unit numbers that have been assigned to active units in the past.

Distinguished Naval Units

There are many notable naval units due to the large number of ships and aviation units currently in place. Some of the more familiar include the "Navy SEALS", the "Blue Angels", the "Seabees", and the Marine "Raiders".

The "Great White Fleet" is probably the most famous, as it contained 16 Battleships that travelled the globe from 1907-

1909 under the direction of President Theodore Roosevelt, who coined the famous phrase; "*Walk softly, but carry a big stick*".

Today, the U.S. Navy consists of seven active "Fleets", encompassing the globe. They are based in the following locales:

======== 1st Fleet - ** [reclassified as the 3rd Fleet]
North Atlantic- 2nd Fleet - HQ @ Norfolk, Virginia
India/Indonesia - 3rd Fleet - HQ @ San Diego, Calif. **
South Atlantic - 4th Fleet - HQ @ Mayport, Florida
Middle East- 5th Fleet - HQ @ Manama, Bahrain
Europe/Africa - 6th Fleet - HQ @ Naples, Italy
North Pacific - 7th Fleet - HQ @ Honolulu, Hawaii
South Pacific - 8th Fleet - HQ @ Yokosuka, Japan
Central Command 10th Fleet - HQ @ Fort Meade, Maryland (now called the Cyber Command Force. No longer a Fleet).

Note: The U, S. Navy currently maintains 71 nuclear-powered submarines, of which 14 carry ICBM's, + 11 nuclear-powered Aircraft carriers, with two more under active construction, and two others in the pre-construction phase.

Distinguished Navy Admirals

Fleet 5 star (*****) Admiral 4 star (****) Vice 3 Star (***)
Rear [lower] 1 star (*) Rear [Upper] 2 star (**)

William D. Leahy [1875-1959] **Fleet Admiral *****
Joined the U.S. Navy in 1893, and a graduate of the Naval Academy in 1897. William fought in the Spanish-American War, the Philippine war, the Boxer rebellion in China, the so-

called Banana wars in Central America, & then in WW II became Fleet Admiral. He became the Chief of Naval Operations in 1937, organizing and supervising the expansion of the U.S. Naval fleet in anticipation of WW II. He retired in 1938, and was then appointed Governor of Puerto Rico in 1939-1940 and then named Chief of Staff to the President from 1942-1949. **Admiral Leahy became the most influential military leader of WW II**. He died at age 84.

Ernest J. King [1878-1956] **Fleet Admiral ******* Graduated 4th in his class of 1904 at the Annapolis Naval Academy. During his career, he commanded Destroyers, Battleships, and Aircraft carriers. At age 42, he was the youngest to ever command a Naval warship at that time. He created the Naval ship numbering system, the Naval ship organizational manual, and developed the process that resulted in the first salvage of the sunken submarine S-4. He was considered a master of operational efficiency.

Admiral King served as a staff officer to Vice Admiral Henry T. Mayo; Commander of the Atlantic during WW I, and King himself became **Fleet Admiral of the Atlantic during WW II** under Admiral Leahy. He was Chief of Naval Operations in 1942. He retired December of 1945 and died in 1956 at age 78.

William F. Halsey Jr. [1882-1959] **Fleet Admiral ******* Graduated from Annapolis Naval Academy in 1904. He commanded Destroyers during WW I, and became a Naval aviator in 1935. He was a strong proponent of Aircraft Carrier-led Task forces. He died at age 78.

As Vice Admiral, (2 **) he commanded the 3rd Fleet in 1940, and was in the Battles of Leyte, Guadalcanal, and Marshall and Solomon Islands. He directed attacks on Japan from the U.S.S.

Enterprise and was **named Fleet Admiral of the Navy in 1945** until his retirement. Known as "Bull" Halsey, he had the reputation of being an aggressive tactician, and an inspirational leader. He was awarded the Navy Cross, 4 Navy & 1 Army Distinguished Service Medal.

Chester W. Nimitz; [1885-1966] **Fleet Admiral *****
As a young man, he tried to get into the Army's West Point, but there were no openings, so he applied and was accepted to the Naval Academy at Annapolis. He graduated with distinction in 1905. Stationed on ships in Asia and the Philippines, he was named commander of a submarine in 1910. He then commanded the Atlantic Submarine flotilla, and in 1912, was sent to the New London Shipyards to oversee ship and submarine construction. He became the nation's leading authority on Submarines. Between WW I and WW II, he commanded battleships, cruisers and submarines, was named Chief of Naval Operations (4 ****) from 1945 to 1947.

Following WWII, He transformed our Naval Forces with the advent of new weapons and the utilization of nuclear-powered ships and submarines. Nimitz was the last Fleet Admiral **(5 *****)** of the Navy. He died at age 80.

Hyman G. Rickover (1900-1986) **Admiral ****.** Known as the **'Father of the Nuclear Navy'**, Admiral Rickover was the longest serving member of the Armed Forces, serving on Active Duty for 63 yrs.

Immigrating from Poland to the United States as a 6 yr-old child, he attended the Naval academy in 1918, during WW I. He served as the Engineering Duty Officer on multiple ships, before volunteering for submarine duty in 1929. In 1937, he translated a captured German manual on submarines into English, which became the model handbook for U.S. Submarines. During and after WWII, he was assigned to various Naval Shipyards in Hawaii, and then Okinawa, Japan. He was also assigned to work with General Electric in 1946 to develop a nuclear power plant for submarines. He led the effort to develop the first nuclear submarine, the *Nautilus.*, launched in 1954. He was promoted to Vice Admiral in 1958, when he was awarded the first of 2 Congressional Medals of Honor. He interviewed every prospective captain of a nuclear submarine.

He was promoted to a 4 star **** Admiral in 1973.

In 1983, in conjunction with his retirement, he was awarded the second of his Congressional Medals of Honor. In addition, Rickover had previously been awarded 3 Distinguished Navy Service Medals, 2 Legion of Merit awards, the previous Congressional Medal of Honor, The Presidential Medal of Freedom, the Enrico Fermi award, and numerous naval citations and medals.

He amassed a zero-incident record of nuclear accidents, which was equivalent to 3000 years of problem-free service.

He authored over half a dozen books, and was considered a tyrant by some, and a hero by others. He stated that "*he had little tolerance for mediocrity, and none for stupidity*". Rickover died of a stroke at age 86.

Distinguished Submarine Commanders

Roy M. Davenport (1909-1987) **Rear Admiral *.** A Sub Commander during WW II and **one of the most decorated servicemen in the Navy**. Graduating from Annapolis in 1933, he served on a battleship before attending the Naval Submarine school in New London, Connecticut. He was assigned to the submarine *Silversides*, & after 4 patrols, to the Sub; *USS Haddock*.

He was the only Sub Commander to be awarded 5 Navy Crosses, 2 Silver Stars, 2 Presidential citations, and two (Navy & Marine) Commendation medals. He had eleven successful patrols during WW II & credited with sinking 17 Japanese ships & disabling 10 more. He retired as a Rear Admiral and died at age 78.

Richard Hetherington O'Kane (1911-1994) **Rear Admiral *. The most successful Submariner ever, and also the most highly decorated Naval Officer.** He was awarded the Congressional Medal of Honor, 3 Navy Crosses, 3 Silver Stars, a Legion of Merit, plus many other awards.

His career started onboard on a heavy Cruiser, then a Destroyer, and he then was sent to submarine school in 1938. During WW II he was credited with sinking 33 Japanese ships. His own ships' torpedo malfunctioned, and it sunk his own submarine & he was then captured by the Japanese. Following his release, he was an instructor at the Armed Forces Staff College, then to the New London Submarine School, where he became the Commanding officer. He retired in 1957 as a Rear Admiral. He died at age 83.

Ed Latimer Beach Jr. (1918-2002) **Sub Commander.** He graduated from the Naval Academy and served on several submarines prior to being an advisor to President Eisenhower.

Awarded the Navy Cross for sinking 9 enemy ships in a single patrol, plus 2 Silver Stars, Bronze star, Legion of Merit, and 3 Presidential citations. He also authored two award-winning books including; *Run Silent-Run Deep,* & *Submarine!* He retired in 1966. His father; Ed Beach Sr., was also a distinguished Naval Captain.

William McRaven (1955-living) **4 star Admiral ****.** The son of a career Air Force officer, William attended the University of Texas, graduating in 1977. He was a member of the Naval R.O.T.C. He also holds a Master degree from the Naval Post Graduate school. **He is currently the highest-ranking Navy Seal.**

After graduation, he volunteered for the Basic Underwater Demolition /Seal training. He became a Naval Special Warfare Officer. On 1982, he was assigned to Seal Team Six. He then commanded Seal Team 4, then commander of Seal Team 3.

He was awarded three Defense Distinguished Service Medals, two Legion of Merit, two Bronze stars, plus many more.

Distinguished Naval Aviators

James Bond Stockdale (1923-2005) **3 star Vice Admiral *****

Graduated from the Naval Academy in 1947. He served on numerous ships before becoming an aviator. Upon graduation to pilot, he became a test pilot. In 1964, he was sent to Viet Nam. As a fighter pilot, he was shot down in 1965 & captured by the Viet Cong.

After 7 ½ years, he was released. James was the most Sr. Officer held captive by the N. Vietnamese. He was officially the last serviceman out of Viet Nam. He retired in 1979, after serving as President of the Naval War College. He ran as Vice President of the U.S. with Ross Perot in 1992. He lost, but got 19% of the Nat'l vote. He died of complications of Alzheimer's' at the age of 81.

Distinguished Navy Sailors

David L. Isom (1967-Living) **Navy Seal - Sr. Master Chief** Born in Hendersonville, N.C., David joined the Navy in 1987 at age 20. He was selected to attend basic underwater demolition school [BUDS/SEAL] training, and became a Navy Seal assigned to Seal Team 1, then was assigned to Special Reconnaissance Seal Team 2. He was Promoted to Special Operation Command Pacific, then served in Iraq, Afghanistan, & Kuwait. Holds a BS. degree from Norwich University in Strategic Studies

Currently, he is the senior Enlisted advisor to the Chairman of the Joint Chiefs of Staff at the Pentagon. **He is the highest ranked enlisted man in the Armed Forces.**

Awards: Seal Trident, Paratrooper badge, Defense Superior Service with Bronze cluster, Legion of Merit, 4 Bronze stars, Navy Commendation medal, Marine commendation medal, Defense Meritorious Service medal (2), plus 18 more.

Chapter Twelve

United States Marines

"Semper Fidelis" – Always Faithful.

The Marines were originally established in 1775, during the Revolutionary war. They were disbanded in 1783 following the end of the Revolutionary war, but re-established in 1798, when pirates along the Barbary coast on North Africa began attacking the ships of the Spanish, French, and British. Spain asked newly-elected President Thomas Jefferson for help. He obliged by re-activating the Marines.

The Marine Hymn is about the history of the Marine Corp. Most people do not realize it, but the opening words "From the halls of Montezuma to the shores of Tripoli" recall our history.

The "*Halls of Montezuma*" refer to the Mexican-American war [1847-1848] and specifically the battle of Chapultepec in Mexico City, which was the capital of Mexico and considered to be the halls of the Aztec emperor; Montezuma. When the

U.S. defeated the Mexicans, part of what is now Southern California. Arizona, and New Mexico, became part of the United States.

The "*To the shores of Tripoli*" refer to the capital of Libya, where Arab pirates were continually attacking European ships and holding its passengers and crew for Ransom. When Americans were seized, President Thomas Jefferson refused to pay ransom, after which, Tripoli then declared war on the U.S.A. [1801-1805]. Jefferson then sent in the Marines & the Arabs were defeated.

The Marines Mission

"To fight our nation's battles, wherever they may be, and more importantly, to win every time."

How the Marines are Organized

The Marines are organized similar to the U.S. Army and shares the same ranks as the Army; however, the Marines are officially part of, and are under the command of, the U.S. Navy.

The Marines are divided into four groups of three; 3 Marine Expeditionary Forces, 3 Aviation Wings. 3 Combat Fire teams, & 3 Logistics Teams. Multiple teams make up a Squadron, multiple Squadrons make up a Platoon, multiple Platoons make up a Battalion, and multiple Battalions make up a division.

The Marines are primarily an offensive force as opposed to defensive. When it is time to take a ground battle to the enemy, we send in the Marines. The Marines are a lightly equipped, highly skilled, fast moving, offensive force, while the Army tends to be charged with holding any seized territories and defending our home territory with heavily armed men, & heavier support equipment.

Distinguished Marine Divisions

The 1st Marine Division – based in Camp Pendleton, CA.
The 2nd Marine Division – based in Camp Lejeune, N.C.
The 3rd Marine Division – based in Okinawa, Japan
The 4th Marine Division – Inactive – Wartime only
The 5th Marine Division – Inactive – Wartime only **

** Forever immortalized as it was the 5th Marine Division that raised the flag on Mt. Suribachi on Iwo Jima during WWII

Distinguished Marine Generals

Lewis Burwell Puller (1898-1971). Lewis "Chesty" Puller **is the most decorated U.S. Marine in history.** He was awarded 5 Navy Crosses plus the Distinguished Service Cross, 3 Air medals, 2 Legion of Merit awards, a Silver Star, a Bronze Star, just to name a few. He enlisted in the Marines as a private and then applied to Officer Candidate School (OCS), graduating as a 2nd lieutenant over the Marine Reserve forces. The reduction of forces following WW I, saw him reduced in rank to a Corporal and placed on inactive duty.

He was re-activated and sent to Haiti. Puller served in the Banana Wars in Haiti for five years and was finally reinstated as a second lieutenant in 1924. In 1928, he was sent to Nicaragua. He returned in 1932 and completed a one-year Officer course. He was sent back to Nicaragua, then to China in 1934 to command the Chinese Marines. In 1942, he was sent to defend Samoa, then onto Guadalcanal, where as a Lt. Colonel, he re-organized the forces which led to our victory.

Following numerous victories and garnering more awards, he was sent to Camp Lejeune, where he became Commanding Officer. Then he was sent to command the Marines at Pearl Harbor. During the Korean War, he commanded the 1st Marine division. He retired in 1955 as a Lieutenant General [3 ***]

Distinguished Marine Aviators

Gregory 'Pappy' Boyington (1912-1988) **Marine Aviator**
Born in Cour d'Alene, Idaho in 1912. He graduated from the University of Wasington and joined the Marine Corps in 1935.

He resigned from the Marine Corp early in 1941 to join the American Volunteer Group and served in China to fight Japan. Following the attack on Pearl Harbor, he rejoined the Marines and commanded the Marine Fighting Squadron, knick-named the "Black Sheep" squadron, which shot down over 200 Japanese aircraft in 120 days. Known as the "Flying Tigers", he was credited with 28 "kills".

Captured by the Japanese in 1944, he was held for 20 months. and awarded the Congressional Medal of Honor posthumously, when he was believed killed in action. When he was found alive and released in 1945, he was presented the award, personally.

He retired a **Colonel** in 1947, and started the Flying Tigers Air Cargo business, hiring many former war-time pilots. He also wrote a book about his experiences which were turned into a television series; "Black Sheep Squadron". [1976-1978].

Distinguished Marine Officers

John Ripley (1939-2008) **U.S.M.C. – Colonel.** Born in Radford Virginia, John was one of the most celebrated officers in the Marine Corp; He was awarded the Navy Cross, a Silver star, Bronze star, the Legion of Merit, The National Defense Service, a Presidential Unit Citation and both the Navy and the Marine Commendation medal plus at least 8 more significant awards, and ribbons. Although he has received every imaginable medal and award, he was not awarded the Congressional Medal of Homor, since his actions did not save any lives while accomplishing them.

He was one of only a handful that had accumulated all four Special Operations Combat skills. He was a certified SCUBA instructor, an Underwater Demolition Expert, a Paratrooper, and both a rifle and pistol Expert. He was best known for his single-handed demolition of a Vietnamese bridge while suspending himself under the bridge for three hours in order to plant explosives all while under enemy fire. The destruction of the bridge prevented the North Vietnamese from moving supplies south into South Viet Nam.

He died at age 68, and is buried in the Annapolis Naval Cemetery.

James H. Webb (1946- living) **USMC – Colonel / Senator** Born in 1946, in St. Joseph, Missouri, the son of an Air Force pilot. James enrolled in the U.S. Naval Academy, and graduated first in his class at the Marine Officer's Basic School.

As a member of the 5th Marine Division, James was sent to Viet Nam. While there he was awarded the Navy Cross and 2 purple hearts for injuries which ended his military career in 1972.

He then enrolled in Georgetown University, becoming a lawyer, and was named Minority counsel to the House of Representatives on Veteran Affairs. Then named as Assistant Secretary of Defense for Reserve Affairs, and then Secretary of the Navy. He ran for political office, was elected Senator, and drafted legislation on the post 9-11 G.I. Bill, which became law.

As the recipient of more than two dozen National civilian awards, James also became a writer, and received an Emmy award for his coverage of Beirut in 1983. In 2004 he was embedded in the military forces in Afghanistan, and was Executive Producer of the film; Rules of Engagement. James H. Webb is now retired.

James Norman Mattis (1950-Living). Born in Pullman, Washington, his father was in the Merchant Marine and his mother, who had worked for Army intelligence during WW II, was working at the plant supplying fissionable material for the top-secret Manhattan Project. James graduated from Central Washington University in 1971. He obtained his Masters' degree from the National War College in 1994.

Mattis served in the R.O.T.C. and then joined the Marines in 1972. He initially served in the 3rd Marine Division, then as a part of the Naval Academy Preparatory battalion, He went on to command a rifle company in the 1st Narine Regiment.

As a Major, he managed a Portland, Oregon Recruiting Station, and then as a Lieutenant Colonel, he commanded the 1st Battalion, 7th Marines in the Gulf war. In 1994, he commanded the 7th Marine Regiment. As a Major General, he led Commands in Afghanistan, Iraq, and Iran.

Mattis was named Supreme Allied Commander of NATO. He was layer named Secretary of Defense by

President Trump in 2016. He later resigned in 2019, when he and the President agreed that their respective Leadership styles were not compatible.

James Mattis was awarded the Defense Distinguished Service Medal, Navy Distinguished Service Medal, Legion of Merit, Bronze star, Combat Action Medal, plus 30 more awards.

Troy Black (1969-Living) **U.S.M.C. - Sargeant Major**

He was a Drill Instructor at Paris Island Recruitment Center, who volunteered to go to Iraq and Afghanistan, where he served two tours. While there, he rescued a wounded soldier who was injured by an IED and vulnerable to enemy fire. He was awarded the Distinguished Service Medal, both a Navy and Marine Defense Service Medal, a Bronze star, Legion of Merit, a Fleet anti-terrorism medal. When he returned Stateside, he was named as an advisor to the Joint Chiefs of Staff, & only the 16th Sergeant Major of the Army.

"The Few, The Proud, ... The Marines"

* * * * *

Chapter Thirteen

United States Coast Guard

"Semper Paratus" – Always Ready

The United States Coast Guard was formally established in January of 1915 by President Woodrow Wilson, but it had actually been in existence since 1790 operating as the Revenue Cutter Service.

Alexander Hamilton proposed that the government build ten Cutters to protect the nations' tariff revenue by intercepting smugglers intent on avoiding our import taxes, and preventing illegal immigration by slave traders, which it did.

The Coast Guard, while technically part of the U.S. Navy, is only assigned to operate under the Navy during war time.

The Coast Guard Mission

"To ensure the safety, security, and stewardship of the Nation's waters."

The U.S. Coast Guard is probably the most under-rated service of all the services. The reality is that the Coast Guard has some of the toughest Basic Training of any service except for the Marines.

In order to perform the duties assigned to the Coast Guard, which includes intercepting drug smugglers on the open ocean, evacuating sick or injured seamen or passengers from ships at sea, chasing Russian fishing trawlers who poach our fish from Arctic waters, to rescuing vacationing boaters from disabled or sinking pleasure boats, Coast Guardsman have to undergo and maintain their strength & stamina throughout their C.G. career.

How the Coast Guard is Organized

Like the U.S. Navy, the Coast Guard is divided into two main areas', each consisting of waterway Districts of Responsibility.

The **Atlantic** Seaboard is divided into 5 districts; Boston MA, Cleveland OH, Portsmouth VA, Miami FLA, and New Orleans LA. The **Pacific** is divided into four districts; Alameda, CA, Seattle WA, Honolulu, HI, and Juneau AK. There is also a large Sea-Rescue base on Kodiak Island, under the Juneau district.

Each District is the divided into Sectors. Each Sector is commanded by a Captain. [O-6]. All inland waterways are assigned by Sector. Coast Guard Commandant's are equal to a Navy Admiral.

Distinguished Coast Guard Commandants

Thad Allen [1949-2010]. The 23rd Commandant of the Coast Guard (equiv.=1 star General) graduated from the Coast

Guard Academy in New London, Connecticut in 1971. He also has a Masters' degree in Public Administration from George Washington University, plus a Masters in Management from M.I.T. Sloan Management School.

After commanding several ships, he became Commandant of the Coast Guard 7th District, which includes everything between Virginia and Key West & the entire Caribbean. Then he was Commander of the Atlantic, which includes the Great Lakes, to the Gulf Coast and everything in between. He directed the Federal response to hurricane Katrina, hurricane Rita, & served as the National Incident Response Commander during the Deep Horizon Oil spill.

He concluded his career as Vice Admiral (**4 ******) Chief of Staff at U.S.C.G. headquarters in Washington D.C.

Miles H. Imlay [1902 - 1975] Rear Admiral. (**3*****) Awarded 2 Legions of Merit, and a Silver Star. Born in Washington D.C. attended Brown University, graduated from the Coast Guard Academy in 1926. Commissioned as an Ensign. Served as a periodical instructor at the USCG academy. Deployed to the invasion of Normandy in Operation Overlord. He retired in 1956.

Distinguished Coast Guardsman

Douglas Munro (1919-1942) **USCG – Signalman 1st class**
Douglas was born in Vancouver, Canada. His parents moved the family to Cle Elum, Washington as a child.

He joined the Coast Guard in 1939 and after serving in several South Pacific battles, he was sent to Guadalcanal. He was in charge of 8 "Higgins" boats transporting and shielding a landing craft that was evacuating Marines, whom had been over-run by the Japanese.

Using a machine gun, Munro used suppression fire to minimize enemy fire, which allowed the remainder of the troops to be extracted, before he himself was shot. His last words were "Did they all get off?" He died September 1942. He received the Presidential Medal of Honor posthumously, and is the only Coast Guardsman to ever receive this highest military honor.

Bernard C. Webber (1928-2009) **USCG Chief Warrant Officer.** In February of 1952, Cape Cod, Massachusetts was being pounded by a Nor'easter. An oil tanker, known as the *Pendleton* was slowly making its way through the storm, which had generated waves reported to be 60 feet high. The high waves pounded the side of the 500-foot-long ship with such force that it split the hull at the seams. As the ship radioed an S.O.S. it began to sink with 33 men onboard.

The nearby Coast Guard station at Chatham, responding to the call for help, launched a 36-foot rescue launch into the freezing rain and the treacherous water conditions. Petty Officer; Bernard C. Webber was the Coxswain and the man in charge, along with three other Coast Guardsmen.

By the time the launch reached the ship, it had broken in two, trapping the ships' crew on the Stern section. One crewman had been trapped in the flooded bow section and had drowned. The remaining 32 crewman fled to the Stern and were all rescued, and returned safely to Chatham.

Bernard, and his three crewmen were each awarded the Gold Lifesaving Medal. Bernie went on with his Coast Guard service and served in Vietnam before retiring as a Chief Warrant Officer 4, following his 20-year career.

That rescue is still considered the most daring and heroic rescue in Coast Guard history.

Nathan Bruckenthal [1979-2004] **USCG Fireman 3rd class** Awarded the Armed Forces Expeditionary medal, the National Defense Service medal, the Global Terrorism medal & a Bronze star for combat bravery

A volunteer fireman from Connecticut, he joined the Coast Guard in 1999. He served in Washington State, then Florida, before being assigned to the Persian Gulf. He served two deployments there. He was a Damage Control (fireman) 3rd class.

He was killed in action in the Iraq war, when he and two sailors intercepted an Iraqi suicide attack on an offshore oil terminal in the Persian Gulf. He was the first and only Coast Guardsman killed since the Viet Nam war.

The Coast Guard at War

The most severe war-time action undertaken by the Coast Guard resulted in the loss of 18 men at Guadalcanal during WW II. Signalman Douglas Munro was posthumously awarded the Congressional Medal of Honor for his actions there.

The greatest loss, was of 18 men, and 38 wounded during landing at Normandy on D-Day. It represents the single largest number of casualties in a single action.

"War time in the Coast Guard, is every day".

The Surf Men

Have you ever heard of the Surf men? Few people have, but these 130 or so highly skilled Seaman who practice rescue techniques by rolling their patrol boats upside down into 30-foot waves, and then recover back to upright, as they cross the Columbia "Bar" off the coast of Oregon, which just happens to be one of the most treacherous waters a ship, or pleasure boater will ever encounter.

Did I mention the sharks? By contrast, Navy Seals do perform some of the most dangerous jobs in the military, but I doubt they would volunteer to do what the Surf men practice every week.

* * * * *

Chapter Fourteen

United States Air Force

"Semper Sursum" - Aim High

The United States Air Force was created as the Aeronautical division of the Army Signal Corp in 1907. Over the years as aircraft technology and capabilities improved, the name was changed in 1918 to Army Air Service, then to the Army Air Corps in 1926, then to the U.S. Army Air Force in 1941.

Famed General "Billy" Mitchell was the Deputy Director of the Air Service, and following WW I, argued profusely that the Army and the Navy were not utilizing the potential of air power, and argued that the Air Force should be run as a separate military service. He touted that a single airplane could sink a battleship, and a test was set-up, in which a lone airplane did indeed, sink a foreign naval ship seized in WW I. The point was well-made and taken.

Following WW II, having proven its value, President Truman was convinced to act. and finally, with the adoption of Trumans' National Security Act of 1947, the Air Corps separated from the Army, and is to today, simply the U.S. Air Force.

Following WW II, General Carl A. Spaatz became the first Chief of Staff of the Air Force, becoming equal to the Army and Navy as the third leg of our military Triad of Defense.

The Air Force Mission

"Fly, Fight, Win.... Airpower anytime, anywhere".

How the Air Force is Organized

Structurally, the Airforce is broken into five functional organizations within the United States, and eleven geographically elsewhere. This results in sixteen (16) air command forces reporting to Headquarters, which consists of four key command organizations;

Air Combat [ACC] **Transport** [AFMC],
Reconnaissance [AFSOC] **Training** [AETC] forces.

Numbered Forces – USA

1st Northeast USA
2nd Northwest USA
3rd Southeast USA
4th Southwest USA
5th West/Central USA

Numbered Forces – Global

6th Philippines/Central Pacific
7th Hawaii/Western Pacific
8th Europe/Central
9th Middle East/N.E. Africa
10th India
11th Alaska/North Pacific
12th Mediterranean/W. Africa
13th South Pacific
14th China/SE Asia
15th Mediterranean
[no 16,17,18, or 19] 20th Marianna's/Indonesia

The Air Force Space Command [AFSPC] is now part of U.S.S.F. The Command structure generally aligns with the following: Forces, Wings, Groups, Squadrons, Flights, & Elements.

Air Force Commands

The Strategic Air Command was headquartered in Nebraska., It was dis-banded in 1992, and replaced by the Global Strike Command (GSC), the Air Combat Command (ACC), the Air Mobility Command (AMC), and the Air Education and Training Command. (.AETC) These four commands are broken down into 9 main Air Force Groups and 7 Global Geographic groups.

1st Air Force

The 1st Air Force is headquartered at Tyndall Air Base, near Panama City, Florida, and falls under the ACC. It is responsible for the defense of the contiguous U.S.A, Puerto Rico, and the Virgin Islands.

2nd Air Force

The 2nd Air Force is headquartered at Keesler Air Base in Biloxi, Mississippi. It manages all of the basic military training and technical training in 256 specialties. It is under the ACC.

3rd Air Force

The 3rd Air Force is headquartered at Ramstein Air Base near Kaiserslautern, Germany. Its mission is to serve NATO countries, plus Europe Africa & Middle East as part of the ACC.

4th Air Force

The 4th Air Force is a Reserve command that operates from March AFB, in Riverside, CA. It provides airlift capability for the U.S. west coast. It is part of the Air Mobility Command (AMC.).

5th Air Force

Established in 1941, the 5th Air Force has ruled the Pacific Ocean for more than 80 years and is based in Yokota Air Base in Japan. It is under the Air Combat Command (ACC). Some components are also based in S. Korea.

9th Air Force

The 9th Air Force is responsible for Latin America, which includes Central & South America, the Caribbean. It is headquartered in the Panama Canal Zone, Panama. Components are under the Air Mobility Command (AMC).

10th Air Force

The 10th Air Force is a Reserve Force headquartered in Fort Worth, Texas at a joint base operated by the Navy. Its primary mission is tactical support for Special Operations & the United States Space Force. (U.S.S.F.)

15th Air Force

The 15th Air Force is headquartered at Shaw Air base near Sumpter, South Carolina. It manages the Reserve & National Guard flight training for the Air Combat Command. (Air Force Reserve Command AFRC)

16th Air Force

The 16th Air Force is based at Joint Base San Antonio at Lackland Air base, Texas. It is the focal point of Intelligence gathering, Cyber Operation, Weather, Surveillance, Electro-magnetic warfare (ACC).

20th Air Force

The 20th Air Force is responsible for the management & deployment of the nation's ICBM missile defense system. It is headquartered in Cheyenne, Wyoming. It is part of the Global Strike Command (GSC).

22nd Air Force

The 22nd Air Force is a Reserve Mobility Command under the (AMC). It is headquartered at Robins Air Base in Warner Robins, Georgia. It provides airlift capability to the East coast

Distinguished Air Force Generals

The U.S. Air Force follows a hierarchical Command structure similar to the Army, which is not surprising, given its history. Generals are designated 1,2,3, or 4 stars, followed by Colonels, Majors, etc.

James Harold Dolittle [1896-1993] Four star ** General**

Although born in California, James Doolittle grew up in Nome, Alaska. He joined the Army in 1917, and moved into the Army Air Corp Reserves in 1918. He was commissioned as a second Lieutenant, serving as a WW I flight instructor.

Graduating from U.C. Berkley, he was a talented pilot. He made his first cross country flight in 1922, and was awarded a Doctorate degree from M.I.T. in 1925. In 1929, he made the first all-instrument flight from Florida to California, which pioneered all-weather flying. In 1932, he set the speed record for flight at 296 MPH. becoming the fastest man in the world.

He was activated from the Reserves in 1940, and in 1942, he planned and flew the first B-25 long-range, top secret, bombing attack on Tokyo. His repeated long-range attacks resulted in his Air Wing being nick-named Doolittle's Raiders.

He was a Lieutenant General **, and commanded the 12th Air Force over North Africa and then the 15th Air Force over the Mediterranean, & then the 8th Air Force over Europe, where he dramatically changed U.S. air tactics. He then reverted from "active duty" to "reserve status".

He became assistant Chief of Staff to the President in 1952, and in 1958, became the first head of NASA, as he was a strong proponent of rocketry. He officially retired in 1959. His sons; James Jr, and John both became fighter pilots, with John becoming a Test pilot at Edwards Air Force base.

General Doolittle was promoted to honorary Four star **** General in 1985. He was awarded the Congressional Medal of Honor, two Distinguished Flying crosses, a Silver Star, a Bronze Star, the Presidential Medal of Freedom, & Distinguished Air Medals from four foreign governments.

Curtis Emerson LeMay [1906-1990**] 4 star (****) General** Born in Columbus, Ohio in 1906 and graduated from Ohio State University with a degree in Civil Engineering. Upon graduation, he joined the Army Air Corps as a Cadet at the age of 22. In 1929, he became a fighter pilot, then cross-trained into bombers. He helped develop air routes across the Atlantic & South Pacific prior to World War II.

During the War, he led the 305th Bomb Group, bombing in Europe and then Japan. He flew the lead plane in every bombing raid he planned. He ordered and led the fire-bombing of Japan which killed an estimated ½ million Japanese civilians, prior to the U.S. dropping the atomic bomb.

He developed the concept of a Strategic Air Command; an all-jet long-range bombing fleet. At age 55, he became Air Force Chief of Staff (1961), Nominated for 5-star status in 1965, but was turned down because this was a war-time rank only.

He ran as an Independent Vice President candidate with George Wallace in 1968. He claimed that President Johnson was too weak on bombing the enemy in Viet Nam. He got

13.5% of the national vote. In time, LeMay's war theories proved correct.

LeMay was the youngest serviceman to ever reach four-star General (age 44). He was also the longest serving General (1941-1965). He was known as "Old Iron Pants". He was revered as an innovator, and a tireless, ruthless leader.

Distinguished Air Force Aviators

ACE pilots are defined as those shooting down fifteen or more enemy aircraft in air-to-air combat. Eddie Rickenbacker was the acknowledged first ACE of our armed Air Forces with 28 "kills".

ACE status was raised to 25 during WWII, and then to 30.

Defensive technologies make shooting down opposing aircraft much more difficult today. The new ACE minimum is five "kills". Only five American pilots achieved ACE status during Viet Nam, compared to roughly 1300 ACE WW II pilots, and 350 during WW I.

Francis "Gabby" Gabreski [1919-2002] **Lt. Colonel.** Born in Oil City, Pennsylvania of Polish parents, Francis attended Notre Dame. He became interested in flying, and volunteered to join the Army Air Corp in 1940. He earned his wings in 1941 and was commissioned a Second Lieutenant.

While stationed in Hawaii, he read about the successful Polish squadron within the Royal Air Force. Being of Polish heritage, he proposed that the U.S. could learn of their tactics and incorporate those into U.S. tactics. He was sent to England, and was a quick study, flying 20 missions with the polish squadron.

In 1943, taking command of the U.S. 61st fighter squadron in France. By March of 1944, he had 18 victory credits and 6 personal "kills". By July, he had become America's leading "Ace" with 27 total "Kills". He crashed on his final run before being shipped home and was captured by the Germans.

After his release at the end of the war, he went to work for Douglas Aircraft, but was recalled to duty in the Korean war.

During his first seven months of duty, his squadron scored 96 MIG "kills". He himself, became an ACE once again. He was one of only seven U.S. pilots to become an ACE in two wars.

Gabreski was awarded the Distinguished Service Cross, a Silver Star, a Bronze star, Legion of Merit, a P.O.W. medal. And Distinguished Service medals from 5 foreign governments including Poland. He is now ranked #2 Ace of all times [28 kills].

Robert Samuel Johnson [1920-1998] **Lt. Colonel.** Born in Oklahoma, Robert worked in his fathers' cabinet shop after school. He applied his meager earnings toward flying lessons at age 12. At age 21, he enlisted in the Army Air Corp, and in 1942 began basic flying training at Randolph field in Texas, where he was commission in June of that year.

In July he was transferred East to the 61st Fighter wing in Connecticut, then to New Jersey, and then in January of 1943, to The 56th fighter group in Scotland for additional training.

By November of 1943, the 56th had flown 89 missions, and had amassed a string of victories. Robert had surpassed Eddie Richenbackers record of 25 "Kills", and was now being sent stateside. (Robert is now ranked #3 of all Air Aces with 27 kills.). Robert retired in 1962, as a member of the USAF Reserves. He died in 1998 at the age of 78. He was awarded the Distinguished Flying Cross, a Silver Star, a Bronze star, & 3 Air Medals.

Robert "Robin" Olds Jr. [1922-2007] **Brigadier General** At the age of 12, he declared that he intended to go to West Point, which he did in 1940. He graduated in 1944, after also exhibiting a successful Football talent on the West Point team.

Due to the war, the graduating requirements changed. Robin now had a modified curriculum and entered flight school in 1942. He was sent to various air training schools in Texas, California, & New Jersey before being sent to Scotland, then England, and then France in 1944. He initially had five approved "Kills" over Germany in his P-38. He the achieved 8 more "kills" in a P-51, being the only WW II pilot to achieve ACE status twice with two different aircraft.

He was promoted to Lt. Colonel in 1951. He became an instructor, and returned to flying in 1955. After a stint with NATO, he went to Southeast Asia in 1966, where he commanded the 8th Fighter Wing.

As Commander he planned several raids to lure MIG fighters into a trap that would destroy them by air-to-air missiles, plus normal F-4 armaments. He again achieved ACE status, being the only triple Ace Airman in USAF history. His "kills" total 17. Two of his protégé's were Jeffrey Feinstein & Richard Ritchie.

Robin received the Air Force distinguished Flying Cross, Air Force Cross, Silver Star, Legion of Merit, and 22 others.

Jeffrey S. "Fang" Feinstein [1945-Living] **Lt. Colonel**
Born in Chicago, Illinois. In 1963, at age 18, he entered the Military Academy Preparatory School, and in 1964 entered the USAF Academy in Colorado Springs, CO., graduating in 1968.

Following graduation, he was sent to Udorn Royal Thai Air Force Base in Thailand. As a member of the 432nd Reconnaissance Wing, he was part of the 13th Tactical Fighter Squadron, where his actions earned him the Distinguished Flying Cross, and 4 Silver Stars for his first four "kills" and 5 Bronze stars, plus the Air Force Cross for his fifth "kill" in 1972. He was awarded the Legion of Merit & 23 other awards including the Air Force

Commendation medal. Jeffrey later flew AWACs flights during Operation Desert Sheld, & Desert Storm from 1990-91. He retired from the Air Force after serving 33 years.

Richard "Steve" Ritchie [1942-Living] **Brigadier General** Richard was a key sports star in High School and was known as a highly competitive, over-achiever.

In 1964, he graduated from the Air Force Academy with a B.S. degree in Engineering. He trained as a pilot in Texas and in Florida, where he trained on several types of aircraft. He was assigned to the 480th Tactical Wing at DaNang in S. Vietnam, in 1968, where he completed 195 Combat missions.

In 1969, he became an instructor at Nellis Air Force Base in Nevada. Richard volunteered for a second tour in Viet Nam in 1972, and was sent to the Thai Air Force Base in Udorn, Thailand. He flew F-4's with the famed 555th wing.

In May 1972, he got his first two "kills", with two more "kills" in July. His fifth official kill took place in late August.

Richard Ritchie completed 339 combat missions before leaving active duty, & joining the Air National Guard in 1974, then into the Air Force Reserves in 1985. He was promoted to Brigadier General in 1994. He retired in 1999 after nearly 40 years.

By virtue of his retirement date being after Jeffrey Feinstein, he laid claim to the title of the last USAF ACE. Richard had accumulated the Air Force Cross, two Distinguished Flying Cross medals, a Silver Star, an Air Medal, Viet Nam Air Gallantry Cross, plus at least ten other awards.

* * * * *

Chapter Fifteen

United States Space Force

"Sempra Supra" - Always Above

The United States Space Force was so-named because our Space exploration business was getting to be a mish-mash of individuals and organizations from the Air Force, the Navy, the Coast Guard, and the National Aeronautics and Space Administration (NASA). There needed to be a command structure to manage budgets, launch priorities, satellite positioning, and Space Shuttle recovery, just to name a few. By creating a single Structure, both management and operations have a hierarchy of mission priorities & accountability.

Today, the Space Force (USSF) can plan, direct, and execute the various demands for communications, military surveillance, national security, and political verification of world events in addition to the N.A.S.A. exploration & analysis of our solar system and beyond.

The Mission of The Space Force

"To protect U.S. and Allied interests in space and provide security capabilities to the joint forces' assets".

How the U.S. Space Force is Organized

The U.S. Space Force is organized under the direction of Department of the Air Force. The top administrative positions share the same military rankings as the Air Force, Army, & Marines. This makes communication between the services easier at each respective authority level.

The head of the USSF is the Chief of Space Operations, a four-star General, with his subordinates fulfilling similar roles to their Armed Forces counterparts. He reports to the Secretary of the Air Force. Many managers are civilian scientists & analysts.

Administrative, legal, and medical support is provided by the Air Force and is led by civilian Patricia Mulcahy, a retired Army Colonel, who is the U.S.S.F. Deputy Chief. Many members of the U.S.S.F. hierarchy are women.

The Space Force is organized into three branches, led by the Chief of Space Operations (CSO) who is located at Headquarters USSF; Huntsville, Alabama. Each branch consists of Field Commands, which are made up of two or more Delta's, and Delta's are made up of two or more Squadrons. Squadrons consist of 2 or more Flights, and each Flight consists of Sections or Elements (Specialties).

Field Commands are essentially individual Missions. Space Force service members (Element members) are referred to as "guardians" rather than soldiers or Airman.

Much of the work being done by the U.S. Space Force is classified, and is separate from the Armed Services work provided by the Intelligence community.

Enlisted personnel have different insignia's than the other services, each reflecting the unique functions of the Space Force. All enlistees must pass the **A**rmed **S**ervices **V**ocational **A**ptitude **B**attery [ASVAB] entrance test.

- **Space Operations Command**

    David N. Miller Jr. – Lt. General / 3 star -
    Space Operations Commander

Deanna M. Burt – Lt. General / 3 star **–**
Deputy Chief of Staff

- **Space Systems Command**

    Phillip A. Garrant – Lt. General / 3 star –
    Space Systems Commander

Andrew S. Menschner – Colonel
Deputy Commander – Space Systems Command

Michelle K. Idle – Brigadier General / 1 star – Mobilization Assistant to the Commander

- **Space Training & Readiness Command.**

    James E. Smith – Major General / 2 star
    Space Training Commander (STARCOM)

Matthew S. Cantore – Brigadier General / 1 star
Deputy Commander – Space Training (STARCOM)

Space Force Commanders

Dr. Troy E. Meink – Secretary of the Air Force
A civilian who began his career as an R.O.T.C. cadet. He received his Masters and a Doctorate in Aeronautical Engineering at Ohio State University. He was a KC-135 Tanker navigator in the Air Force, then instructor, then lead engineer for ballistic missile test vehicles. The Secretary of the Air Force oversees the Space Force & the Air Force.

B. Chance Saltzman - Chief of Space Operations (CSO)
Four-star (****) General. He started his career in the R.O.T.C. and joined the Air Force as a "Minute Man" Missile Launch officer, and a satellite operator for the Aerospace Data facility in Colorado. He has been a recipient of the Air Force Distinguished Service medal, the Defense Superior Service medal, the Legion of Merit, plus 20 more.

A career Missile and Space operations officer, he previously Commander of 460th Operations Group, served as the Deputy Chief of Space Operations, Cyber Operations, and Nuclear operations. He was Deputy Director of the National Reconnaissance Office and as Under Secretary for Space.

Michael A. Guetlein – Vice Chief Space Operation Four-star (****) General. He was the first commander of Space Systems, and second Vice Chief of Space Operations. Previously served in the National Reconnaissance Office, the Remote Sensing Systems Directorate, and commanded the Rapid Reaction Squadron.

He has been awarded the Air Force Distinguished Service medal, the Defense Superior Service medal, Defense Distinguished

Service medal, and 2 Legion of Merit awards, plus 10 more awards.

Shawn N. Bratton – (1968-living) **Chief strategist for the USSF.** Four-star General (****). A Former **Arizona Air National Guardsman,** a Tech SGT who graduated from Officer Candidate

School., graduate of both the **USAF** advanced weapons school. and **Naval War College.** Served in Iraq, commanded the Cyberspace Weapons Command, V. Chief of Space Operations.

Gregory J. Gagnon (1972-Living) **Deputy Chief Intelligence Operations.** Major General (***). Served in Afghanistan and in Korea. Greg has an extensive background in Intelligence and Cyberspace operations. He commanded the 67th Cyberspace Ops group, the 495th Expeditionary Intelligence Squadron, 94th Intelligence Group, & the Texas Cryptographic Center. Awarded the Defense Superior Service Award, 3 Legions of Merit, & Bronze star.

Steven P. Whitney – (1993-Living)**Director of Staff U.S.S.F.** Major General (**) 2 star – Previously the military deputy to the Assistant Secretary of the Air Force. Steven graduated with a B.A. in electrical engineering and holds a master in in Administrative Services, a Master of Science, and a graduate of the War college.

He has been awarded the Defense Superior Service Medal, the Air Force Distinguished Service Medal, and 2 Legions of Merit.

* * * * *

Note: Each military service has a different ASVAB (Armed Services Vocational Aptitude Board) entry minimum.

The Army has the lowest minimum entry overall, and the Coast Guard, the highest overall. Some specific positions in each service may require a much higher minimum ASVAB score. The Space Force has the highest score threshold for any single position.

Chapter Sixteen

National Guard & Reserve Forces

"Semper Paratus" - Always Ready - Ready Now!

The National Guard and the Military Reserve Forces are not the same thing, but initial basic training for both is essentially the same.

The National Guard forces are typically used to respond to natural disasters that occur within your State or in some cases assisting the Guard of a neighboring State.

Floods, forest fires, tornadoes, and earthquakes are some of the more common disasters that the National Guard responds to. The National Guard is under the control of your State governor and as a result, you are more likely to be called up during a catastrophe, as opposed to a military conflict.

The Military Reserve forces, which include Army, Navy, and Air Force Reserves, are controlled by the Secretary of Defense, and are used to supplement our regular armed forces when there are shortages of certain specialists during military conflicts. Reservists, if activated, backfill positions vacated by regular armed forces who may be moved to a conflict zone. Depending on your skillsets, you will be assigned to a location best suited for those skills at the moment.

Both Reserve forces and National Guard forces go through the same basic training, depending on the Service they have enlisted in. Advanced Training is focused on the career skills in the particular service you have chosen to join. Naval Reservists might train onboard a ship, Air Force reservists may train on an air base, and Army reservists may train in an Army maintenance shop. There are literally hundreds of career possibilities between the various services.

Licensed Commercial Pilots can join the Air National Guard, or the Air Reserve forces. They must still fulfill their one weekend per month, and their two weeks in the summer commitment, the same as other reservists, but because many of these Reservists are commercial pilots, there are some accommodations that are made as to flight scheduling, depending on the type of aircraft they are qualified to fly, the type of missions needed to be flown, and the status of their Unit at the time. Several commercial pilots that were laid off, were able to transfer to active duty and keep their flying skills current. Many flew missions in Iran, Iraq & Afghanistan, while others flew supply missions to free regular active-duty pilots flying into combat zones.

The minimum commitment to get into the reserves is six years, which consists of one weekend every month, and two weeks every summer. Basic training is typically eight weeks, while advanced skill training varies by the career choices available in each service. Some highly technical skills might require a full year of technical training, while other technical training may be as short as ten weeks.

On average, most Reserve and Guard commitments require a total of six months of initial active-duty training. After that you will have an obligation of the remaining enlistment time (5 ½ yrs), during which your training skills will be updated by your monthly weekends, and your 2 weeks Active Duty at "Summer Camp".

You may wonder what can you possibly accomplish in one weekend, but the weekends you are on duty, you are fulfilling work assignments that the regular full-time troops normally do every day and this helps improve both training and coordination between regular troops and reserve forces during the week. This keeps work moving, and adds to our training readiness. It also provides a means to pass along new procedures, technologies, & use of new equipment.

It should be noted that promotions in the Reserves are still based on your skill proficiencies and time in grade, so if you are willing to apply yourself, you can improve your monthly paycheck while fulfilling your service obligation.

Reservists and National Guard troops may be activated anytime the regular forces are short-handed, or are in war-zones to backfill needed support positions. While this is rare, it did occur in the Viet Nam war, and for the National Guard during State Emergencies if declared by the Governor or the President, whom is the Commander in-Chief of all armed forces.

National Guard and Reserve Mission

Reserve organizations have the same mission as the regular forces, except that the National Guard also has an additional State obligation to assist local authorities during times of natural disasters, such as flood, forest fires, earthquake's, tornados, hurricanes and civil unrest, by maintaining peace, security and public safety during those events.

How the Guard & Reserve force are Organized

Guard & Reserves are organized the same as their full-time forces.

Distinguished National Guard Personnel

The history of the National Guard goes back to the Revolutionary War days. At that time, our national defense force was limited to the Militia, a loose collection of citizen patriots, loyalists, and colonists. The Militia evolved into what is now, the National Guard. Many historical figures were part of the National Guard including Paul Revere, Ethan Allen, & Patrick Henry. Later members included George Washington, Abraham Lincoln & Harry Truman.

Audie Murphy - Army National Guard (Infantry) **1st Lieutenant. Audie Murphy was the most decorated soldier of World War II.**

Congressional Medal of Honor, 2 Silver Stars, 2 Bronze stars, Distinguished Service Cross, Legion of Merit, Legion of Honor. Plus 3 Purple hearts and numerous battle ribbons.

Despite being a "week-end warrior" national guardsman, Audie Murphy proved the value & the skills that our reserve forces provide.

Commander-in-Chief trivia

Out of the 45 men who have served as Commander-in-Chief and President of our nation, only 13 (28%) have never served in the military. The following Presidents **never** served in the Armed Forces;

John Adams – Federalist
John Q. Adams - Democratic Republican
Martin Van Buren – Democrat
Grover Cleveland - Democrat
William H. Taft– Republican
Woodrow Wilson - Democrat
Warren Harding – Republican
Calvin Coolidge- Republican
Franklin D. Roosevelt – Democrat
Bill Clinton - Democrat
Barak Obama– Democrat
Joseph Biden - Democrat
Donald Trump – Republican [served in R.O.T.C.]

Chapter Seventeen

America's Heroes

America's heroes are not limited to Generals, Admirals, or Fighter pilots, even though that is where the focus usually lands.

Certainly, we have had great leaders and great fighters, and together our country has survived attacks from our adversaries and prevailed over our enemies. But for every high-profile battle and every major conflict, there are hundreds, if not thousands, of examples of individual acts of heroism. Every big victory is made up of many small victories.

Territory is won inch by inch, foot by foot, and mile by mile. Enemies are overcome bullet by bullet and bomb by bomb, and sacrifices are made soldier by soldier.

In the next section I tried to capture the accomplishments of the ordinary soldier, airman, and sailor, who collectively contributed to the big victories by their personal victories and personal sacrifices. Some of these stories are remarkable. For every remarkable story are dozens more that have been lost to time.

I have tried to convey what constitutes courage. What compels people to attempt the impossible? It is that loyalty, that patriotism, and that belief in our system that motivates people to unselfishly seize an opportunity to do what is needed.

When you read the names of movie stars and celebrities who fought in World War II, Korea, or Viet Nam, keep in mind that it wasn't their notoriety that compelled them to act. They acted because they had the confidence, the capacity, and the courage to recognize that they alone could overcome the adversity in front of them.

Bea Arthur was a truck driver during the war. That doesn't sound particularly heroic until you realize that these trucks were often filled with ammunition being transported to the front lines. One errant bullet striking her truck would have blown her to bits. Instead, she made it through repeatedly. It was only years later, that we enjoyed her TV performance in *Golden Girls*, that we have come to appreciate what she did. Every trip was a victory.

Julia Childs, that affable cheerful cook on TV served her country as a Spy for the Office of the Secret Service. So did Marlene Dietrich, another actress, whose performances we enjoyed in the movies for years following the war, but their real-life performances were largely unknown and remain secret.

Read on to learn more about our unsung heroes as well as the more famous.

The Unsung Heroes

Now that you have read about each of the military services and their legendary leaders, I thought I should devote some space to those individuals that were not generals, nor admirals, but just ordinary service men and women who went above and beyond. Their names may not be household names, but their bravery and courage should not go unmentioned. They became legendary in their own right.

Unlike the military officers whose names are etched in history, there are thousands of young men & women whose unacknowledged actions not only made a difference, but in many cases, saved dozens of lives, and sometimes their own as well. I wish there was space for them all. A few of these young men and women went on to make a name for themselves in business or in the movies. Some even became President, like Ronald Reagan.

The important thing is, that the unselfish efforts of ordinary men and women should not be forgotten or ignored just because they never went on to fame or fortune. These are the unsung heroes of conflict. Some are noted below:

Daniel Joseph Daly (1873-1937) **USMC – Sgt. Major.**
His nickname was "*Devil Dog*", and was known as "*The fightingest Marine that ever lived*". One of only two Marines to be awarded 2 Congressional Medals of Honor for bravery. The first was in 1900 in China, and the second in Haiti. A third, at Belleau Woods in WW I, was denied because they felt 3 would be excessive

John Basilone (1916-1945) **USMC** - **Gunnery Sergeant**
Born in Buffalo N.Y, John's family moved to Raritan, New Jersey. He enlisted in the Army at age 17 ½. After serving two years in the Infantry, He quit for one day, and then re-enlisted in the Marine Corps the following day, stating the Army was not tough enough.

He was sent to Cuba, then was later sent to Samoa. Then he was sent to Guadalcanal. He and two others held off the

Japanese for three days using only 2 machine guns and his pistol.

He was then sent to Iwo Jima where he led a tank through a minefield & took out a Japanese blockhouse, which was preventing the Marines from establishing a beachhead on Red Beach.

He was awarded the Presidential Medal of Honor and posthumously, the Navy Cross. He was killed in action in 1945.

John F. Bolt (1921-2004) **USMC – Naval Aviator - 2nd Lt.** John Bolt was born in Laurens, South Carolina. At age 20, he joined the Marine Corps and then signed up for the U.S. Navy Pilot Training program. In 1943, he was commissioned as a Second Lieutenant and sent to Pacific as a fighter pilot on the aircraft carrier *Rochambeau*.

Once in the Pacific, he was placed in a new squadron, which became known as the "black sheep" squadron under the command of Major "Pappy" Boyington. He was sent to the Marshall Islands to escort B-24 Bombers.

On one occasion, he disobeyed orders and flew a solo mission and damaged four enemy supply barges and strafed Japanese land encampments. After shooting down several Japanese fighters in various engagements, he became the squadron's sixth flying Ace.

In 1952, he went to Seoul, S. Korea, flying with a Marine Squadron. He was engaged in dog fights with MIG fighter jets multiple times, & succeeded in shooting down his attackers on each occasion. After Korea, he became an instructor.

John was the first aviator to a attain ACE status in 2 wars. He flew for the Navy, the Air Force, & the Marines and retired as a Marine Lt. Colonel in 1962. He died of Leukemia in 2004.

Harlon Henry Block (1924-1945) **USMC Corporal WW II**. Born in Yorktown, Texas, he joined the Marines along with seven of his classmates in 1943 at age 19.

Henry was one of the Marines who was ordered to go raise a second, larger flag on Mt. Suribachi on Iwo Jima when the first flag was deemed too small to be seen by the Naval ships bombarding the mountain top. He is the immortalized figure burying the pole into a pile of rocks on the mountain top.

Although he survived the event, he was killed later that day by mortar fire. He was originally buried at Iwo Jima, but his remains were later re-interred in Texas. He is immortalized in the bronze statue that has come to signify the turning point of World War II.

Guy Gibaldon (1926-2006) USMC – Corporal. Spoke fluent Japanese, making him invaluable during **WW II,** he continually snuck across enemy lines at night, and convinced the enemy that they would be treated humanely, fed properly, and that resistance was futile. <u>He captured 1,300 Japanese soldiers single-handedly and marched them into camp.</u> He was awarded the Navy Cross, A Siver Star, plus ten other awards.

David Bellavia (1975-Living) **Army – Staff Sergeant Kosovo, Iraq.** Killed 6 insurgents in Fallujah in hand-to-hand combat and close-in fighting, saving the lives of his unit.

Congressional Medal of Honor, Silver Star, Bronze Star, Army Commendation medal, Army Achievement medal, plus more.

Monica Lin Brown (1988-Living) **Army – Private 1st class 2006 Afghanistan** Sent as a Medic to in 2006, <u>Monica was the first woman to receive the Silver Star since WW II.</u> When her convoy came under attack, she braved explosions and enemy fire to treat and save the wounded. Awarded for Bravery.

Lacie "Sonic" Hester (1982-Living) **USAF/ANG Captain 2024 Iran**. Intercepted over 80 drones aimed at her Squadron. Awarded the Silver Star, Bronze star, Army Commendation Medal. National Defense Medal, Reserve Achievement Medal, Plus 14 other combat awards for gallantry in action

---

Hollywood's Real-life Heroes

I am not going to go into the biographical details of Hollywood's men & women who were also Armed Service members, because they already have extensive published biographies that glorify their movie roles, but it is note-worthy that some of our favorite entertainers not only served in the military during WW II, Korea, and Viet Nam, but did so heroically. We admire them for the bravery they portrayed on the screen, but many actually lived those portrayals.

No lights, camera's, re-takes, nor publicity agents, just the courage to put themselves in harm's way for the sake of our country. Their military record says it all.

Audie Murphy - Army National Guard (Infantry) **1st Lieutenant** Audie Murphy was the most decorated soldier of World War II.

WWII. Congressional Medal of Honor, 2 Silver Stars, 2 Bronze stars, Distinguished Service Cross, Legion of Merit, Legion of Honor, 3 Purple hearts & numerous battle ribbons.

James Stewart - **USAF** (Army Air Corps) **Brigadier General.** Highest military ranking achieved by any Hollywood star; **WWII, Korea, Viet Nam.** Bomber Pilot. Numerous medals.

Charles Durning - U.S. Army (Infantry) **Private 1st class** Only soldier in his squad to survive the Omaha Beach landing. **WWII** Silver Star, Bronze star, 2 purple hearts. He took out 7 machine gun nests. As a German POW, he escaped.

Rod Serling - US Army (Paratrooper) **Sgt.** *Twilight Zone* creator. **WW II** Bronze Star, 2 purple hearts, Philippine Liberation Medal

Eddie Albert - US Navy (Reserves) **1st Lieutenant. WWII** Awarded the Bronze Star. Star of TV "*Green Acres*"

Henry Fonda - U.S. Navy (Reserves) Air Intelligence **Lt. JG** **WW II** – Awarded the Bronze star & Presidential Unit Citation

Lee Marvin - **USMC** (Infantry) Sniper **Corporal. WW II** Navy Commendation Medal, Presidential Unit Citation, Combat Action Ribbon, 2 purple hearts, Wounded on Saipan.

Buddy Ebsen - USCG (Damage Control Officer) **Lt. JG** **WWII** - star of the TV show; "*Beverly Hillbillies*"

Glenn Ford - U.S. Navy (Reserves) Public Affairs **Captain** **WW II, Viet Nam -** Navy Commendation Medal

James Garner - US Army (Nat'l Guard) Infantry **Corporal** **WW II & Korea** 2 purple hearts - star of "*The Rockford Files*"

James Arness - U.S Army - **Sergeant** - Star of *"Gunsmoke"* **WW II** - 1 purple heart- wounded in action.

Charles Bronson - U.S.A.F. (B-29 Gunner) **Corporal**
WW II Highest casualty position in the Air Corps. Purple heart

Ronald Reagan - Army Air Corp (Public Relations) **Captain.**
WWII Became 40th President of the United States

Julia Childs - Office of the Secret Service **(C.I.A) Spy**
WW II – star of the first national TV cooking show

Marlene Dietrich - Office of the Secret Service (**C.I.A.**) **Spy**.
WW II – star of many Hollywood drama movies.

Bea Arthur - U.S. Army Ammunition Truck driver – **Corporal. WWII –** star of *Golden Girls* TV with Betty White

Betty White - U.S Army PX Truck Driver for Hollywood – **Private. WWII** – star of *Golden Girls* TV series with Bea Arthur

Johnny Cash – U.S. Air Force Radio operator **Airman 1st class. Korea -** Served only in Germany from 1950-1954.

Steve McOueen USMC Tank Mechanic/Tank Commander Enlisted in the Marines in 1947 after leaving Reform School. Trained as a Tank/Engine mechanic. A **Private*** eventually promoted Tank Commander. An LST carrying tanks hit a sandbar, and sank with men aboard, Steve dived in and saved 5 men, He was then re-assigned to Honor guard for President Trumans' yacht.

A notorious bad boy who was *demoted several times, and spent several times in the Navy Brig for Drunken-ness, fighting, AWOL, disorderly conduct, etc. He was encouraged not to re-enlist.

Clint Eastwood - U.S. Army Swimming Instructor – **Corporal.** Clint was drafted into the Army during the **Korean** war. He was selected to be a swimming instructor at Fort Ord in California. He was to ensure soldiers knew could swim and stay alive in the ocean. He never served in combat, but does have an interesting backstory.

In 1952, Clint was a passenger on a two-seater SkyRaider prop-driven fighter. The training flight developed numerous malfunctions, eventually ran out of fuel, & crashed into the ocean, miles off the California coast. Clint and the pilot, swam three miles to shore. Years later, he learned that the Pt. Reyes area was a Great-White shark breeding area. No medals, because not serving in a war zone.

Elvis Presley – U.S. Army Motor pool Jeep driver – **Sargeant.** Drafted in 1958 served in 3rd Armored Division in W. Germany.

Bob Hope – [Never served in the military]. He volunteered to work for the USO as an entertainer for the military. He was made an honorary veteran by an act of Congress in recognition for his 63 years of entertaining troops overseas every Christmas, world-wide. He died in 2003 at age 100.

Chapter Eighteen

Courage

What is courage? It is the willingness to take a necessary risk to accomplish a higher objective. Taking necessary risks is doing what has to be done. It is putting others ahead of one's self, in spite of the risk. It is unconstrained bravery.

When you read about someone who was later described as courageous, that person invariably says, I just knew that it had to be done and no one but me could do it. From fire fighters to policemen, to the average citizen, we all have the capacity to be heroic or courageous, but it takes more than just capacity. It requires that without a moments' hesitation, you insert yourself into a situation, because you believe you can do something about it, and you know that there is no one else at that moment in time; that has a chance of acting successfully. You risk your life, to save someone else's.

If the military does one thing really well, it trains its members to have the confidence and the courage to effect a change in the situation at great risk to one's self, because in the larger world, you are the one person who holds your own destiny, or that of someone else's, in your hands. It is the confidence in one's own ability.

Bravery is courage in action. Many of the medals, honors, ribbons, and citations state that they are being given for personal courage, valor, bravery, or gallantry. Let me explain the difference, since these awards mean different things.

Armed Forces Service Medals

The first Congressional Medal of Honor was awarded in March of 1863 to Army Private Jacob Parrot, for his actions during the Civil war. He was one of six Andrew's "Raiders" who participated in the "Great Locomotive Chase".

So, what do these medals really stand for? Here is a brief definition of the terms used in awarding these medals and ribbons.

The Congressional Medal of Honor is the nation's highest honor bestowed on a military service man or woman. It requires Congressional approval. It recognizes the highest level of risk, courage, and sacrifice undertaken in the face of extreme adversity and duress, often under enemy fire. It is often awarded posthumously.

The **Distinguished Service Cross** is given when persistent and prolonged courage ultimately lead to victory by overcoming extreme resistance or adversity under enemy fire. Our second highest medal.

The **Silver Star** is the 3rd highest honor a soldier can receive because their actions under enemy fire demonstrated extreme bravery: It is taking action without fear of the consequences or risk to one's self. It is the highest demonstration of both courage and bravery.

The **Bronze Star**, 4th highest medal. Given for extraordinary courage to undertake a high-risk task or assignment in the face of danger to one's self, but not necessarily while under fire from an enemy.

Distinguished Flying Cross [#5] is given for excellence in flight achievement, Can be for wartime or peacetime achievements. If given in wartime, it usually indicative of extraordinary aviation feats like achieving ACE status. In peacetime it can be given to Test pilot for setting multiple aviation records, or Space Force astronauts.

The 6th highest medal is the Defense **Distinguished Service Medal**, and the **Legion of Merit** [7th]**.** They are both awards for Courageous or Meritorious acts depending on the level of risk & the outcome. Usually awarded for gallantry**,** typically, a lone act of bravery, The decision to take personal action despite the risk to one's self.

[The next highest medal could be one of several depending on the acts that preceded the award, and whether it is during peacetime or in a war zone. These include the following]:

National Defense Service Medal. Established in 1953 to recognize active Federal military service during periods of armed conflicts, including terrorism, occurring before December 31st, 2022.

Distinguished Defense Service Medal (each service has its own version). Awarded for providing exceptional meritorious service to society or the government, upon recommendation by a third party.

Soldiers Medal – awarded by the Army for an exceptional act of bravery or heroism in a non-combat situation.

Purple heart – awarded to anyone injured during a military conflict

* * * * *

Chapter Nineteen

Leadership

One of the life-long skills a person takes away from their military service is leadership. Having the confidence, the capacity, and the courage to lead others will pay you many benefits in both your career and your personal life. That confidence, capacity, and courage may save your life and it may save the life of others who are co-workers, service members, relatives, or complete strangers.

When I joined the military, I considered myself an introvert. I wasn't shy exactly, but I typically just went along with whatever my friends were doing, or not doing. As I entered "boot camp", I was initially shell-shocked by the demanding Sergeants trying to get us to do anything, and everything. You quickly learned that if you were spoken to, you had better be prepared to answer, immediately, politely, and loudly, and not just meekly or quietly. You were expected to shout back quickly "Yes Sir", and that you "Heard", "Understood", and were "Acknowledging" the instruction. In the Army, this is often referred to as H.U.A., or "hooah"!

When I arrived at Lackland Air Force base, one of the first things we did was go to an auditorium filled with hundreds of other recruits like me, and it started off with a brief clip of the opening scenes from the movie "Patton". George C. Scott, in perhaps one of his greatest roles, stood in front a large American flag and proceeded to tell us that we had better pay attention, and if we were ever in a battlefield situation, that if you hesitate, if you over-think the moment, you will not survive. You need to act quickly and decisively by shooting the other guy in the belly before he shoots you in the belly. It was a dramatic beginning to my military service.

Given that we were in the midst of the Vietnam conflict in 1966, this message was not easily dismissed even though we had all seen the entire movie prior to our enlistment.

A real General then gave us a welcome speech, then another speech from a colonel, then an excellent feature cartoon on the dangers of drug use, a film on alcohol abuse, a film on venereal disease, and a final send-off from another officer who told us what we had to look forward to in the next eight weeks.

In that first day, in the first three hours, I was exposed to more information and went through several emotions, as I wondered what I had gotten myself into. One thing for sure, I was quickly becoming a man, instead of a teenage boy. The military has no time or patience for cry baby's, slackers, or procrastinators. You grow up pretty quick as you are suddenly expected to know more, and do more. You are expected to think, act, look and speak like a professional soldier, in other words, like a man, or woman who is sure of himself.

During a combat situation, people get killed. There is no time for reviewing job applications, marksmanship scores, or personality tests. If the leader of your team is out of action, the next ranking soldier steps in and takes command. No discussion, no voting, and with no dissension. You have a mission and the mission has to be accomplished, if at all humanly possible. That is the nature of the military and it inspires confidence every minute of every day.

You learn that the best strategy in the world isn't worth the paper it is printed on, if you can't execute the plan properly.

"Improvise if you have to, but never lose sight of your objective".

A typical military response for failing to accomplish a mission or to overcoming an obstacle might be something like this from Patton:

"Don't tell me what you <u>can't</u> do, tell me what you <u>can</u> do!"

When you get out of the military, you will find that you are less patient with people who don't seem to know what day it is, or much less care. Your world is suddenly filled with imbeciles.

That "Sense of Urgency" that you learn in the military is not how most civilian organizations work. That is why the military is much more effective at getting difficult things done in a hurry. Who else could move 100,000 men with all their equipment anywhere on the globe in 24 hours, with only a few hours of advance notice. No other country has that capacity or flexibility.

The Generals and Admirals you read about, earlier in this book, were likeable, but tough, even arrogant. They did not tolerate failure. They expected the impossible, and they usually got it. At times, they were also hard to deal with; argumentative, difficult, stubborn, and sometimes hot-tempered, but they somehow got the job done. That is what leadership is about; Getting it done. Whatever it takes!

* * * * *

Chapter Twenty

Peace Through Strength

Before World War I, Great Britain was the premier global military force. *"The sun never sets on the British empire"* was the reminder to all, that the British were not to be messed with.

The notorious French Foreign Legion was also a global military force to be reckoned with, even though its sphere of influence was primarily in Southeast Asia, the South Pacific, Africa, and the Middle east.

Russia & China had huge armies, but lacked weapon technology, capacity, and in some cases, competency.

Spain, which had once been a world power had relinquished power in most of its territories either through distance or defeat.

The United States, Canada, and most of Central and South America were content to simply focus on their home territory, or neighboring spheres of influence.

Germany and its Prussian satellite countries changed everything. The Ottoman empire, which controlled most of the Balkans, collapsed with the assassination of Franz

Ferdinand, the Archduke of Austria in 1914. This was the single event that preceded WWI.

Russia and Germany moved quickly to exercise authority over the leaderless countries, and by the time the world realized it, Germany was on the move to conquer and seize even more territory. It was expanding its influence through-out Europe and specifically looked at France and Great Britain as potential additions. Disagreements led to conflicts and those conflicts ultimately led to war.

This was a wake-up call to Great Britain, and to its allies like Canada and the United States. We were all ill-prepared for war.

Separated by the Atlantic Ocean gave us the time and protection to gear-up and man-up, to ensure that the Germans were held to the European continent.

Winning the first World War made all of the affected countries face the fact, that more money had to be spent on their military defensive forces. They needed to maintain an armed force large enough to deter the future ambitions of any other adversary that might get the same idea.

Ten years later, Adolph Hitler began his political rise and insisted that Germany should rise again and take back those territories lost in the first World War, and to ignore the restrictions that the Treaty of Versailles had put on German manufacturing and shipbuilding. By 1938, Hitler had used the prior decade to re-arm and plan out the future re-taking of the Bavarian States that had previously been under German control. Hitler invaded Poland in September of 1939.

Once again, the Allied countries were caught unprepared and ill-equipped to stop Hitler's advance, and once again disagreement, led to a conflict which led us into WW II.

After having learned our lesson in two World Wars, the United States was determined that we would not get caught napping ever again. The Korean conflict was quickly responded to, and after brutal fighting, an Armistice to end hostilities was reached.

When Viet Nam was invaded by the Chinese, we again responded, only to find that the Chinese had built up an incredibly large fighting force that was not interested in an Armistice or anything short of victory. Hostilities finally ended when the U.S removed all its troops after the fall of Saigon. This led us to the conclusion that our defensive capabilities were not enough. We had to have an armed force that was competent, and was committed to victory, and that we had the capacity to sustain any engagement for as long as it takes.

The term "Peace through Strength" popularized by President Reagan in 1984 was initially ignored by our adversaries, and some Presidents subsequently gave only 'lip service' to the concept. None the less, the Pentagon paid attention.

It was finally characterized by General Schwarzkopf's "warrior" mentality. A philosophy that saw us totally defeat the Iraqi Army invasion of Kuwait in 100 hours (just over 4 days).

Today, advances in technology can result in a devastating missile attack on a country within just minutes. There is not time to negotiate, you either respond in kind, or perish. The United States is lucky, in the sense that our form of government encourages innovation, inspires ingenuity, and rewards success. It does require courageous leadership, however.

The Atomic Bomb

For those that think the United States was wrong to drop the atomic bomb on Japan, and believe we should apologize; I leave you with the following reality:

If Germany had perfected the Atomic Bomb first, ...They would have used it.

If Japan had perfected the Atomic Bomb first, ...They would have used it.

If Russia had perfected the Atomic Bomb first, ...They would have used it.

The United States did perfect the Atomic Bomb first, ...and we used it.

[Currently, nine countries possess a nuclear bomb].

-

"Make no mistake, if any adversary of the United States thinks that it has achieved, or acquired the perfect weapon system, and they believe that there is an opportunity to overcome our ability to immediately retaliate, they will use it. A sad fact of man's humanity.

500,000 civilians died in Japan, as a result of our use of the atomic bomb. A tragedy to be sure, but critics lose sight of the fact that if Japan had developed the bomb first, they would most assuredly have used it on our civilians, resulting in losses of a like amount here.

Post-war interviews with the Japanese confirmed this strategy. So, while regretful, there should be no guilt. Our actions did end the war, saving hundreds of thousands of both U.S. and Japanese that would not have survived a full land assault. Only a strong, and committed military can discourage an adversary's instinct to conquer".

Peace through Strength is the foundation of our defense

* * * * *

Author Commentary

Considering how many Americans do not serve in our military, you wonder at times, how we are able to sustain ourselves as a free country? Fortunately, our military has demonstrated over and over, that despite the reluctance of some, it still attracts the best and the brightest of our nation.

As you read about the heroism, the bravery, and the courage of our military leaders, you realize that they blossomed in the military because of its culture, its tradition, and its training. It allowed these men to innovate, to strategize, and to execute in ways that led to victory, often despite overwhelming odds.

Admiral Halsey left one of the greatest military quotes of all time.

During the World War II Battle of Leyte, a fellow officer recognizing that they had just beat back the Japanese combined air and sea attack despite overwhelming odds, remarked that our airman, sailors, and soldiers were "great men".

Admiral Halsey responded with these words;

"There are no great men. There are only great challenges,
….. which some ordinary men are compelled to accept"

If you enjoyed this book, please leave a review on your favorite bookseller's website, or your favorite social media site. Thanks!

Other books by Charles ODonnell

The Wilhelm Legacy

The true account of the worst ocean liner disaster in history as told to me by a survivor of that terrifying WWII experience.

Five Amazing Men

Stores of the 5 ordinary men who inspired me. A machinist, a teacher, an executive, a retail supervisor and a tradesman.

The Ultimate Gun Control Solution

Answers, not Arguments. Over two dozen ways to reduce gun crimes without infringing on your second amendment rights.

Buying Cars Ain't Easy

An entertaining look at the authors personal experiences buying & selling over 50 cars. Tips for prospective car buyers.

Read to Succeed,

Great Readers make Great Leaders

The importance of reading in developing your intelligence, enhancing adult management skills, and improving memory.

Managing your I.R.A. made easy

Improve the financial returns from your IRA without adding any more money, by being able to analyze mutual funds in seconds without a calculator or any special financial knowledge.

Made in the USA
Monee, IL
12 November 2026

713c9404-f517-4bf3-82f3-27a8ba39cb1cR01